Causes and Environmental Implications of Increased
UV-B Radiation

ISSUES IN ENVIRONMENTAL SCIENCE AND TECHNOLOGY

EDITORS:

R. E. Hester, University of York, UK
R. M. Harrison, University of Birmingham, UK

EDITORIAL ADVISORY BOARD:

Sir Geoffrey Allen, Executive Advisor to Kobe Steel Ltd, **A. K. Barbour,** Specialist in Environmental Science and Regulation, UK, **N. A. Burdett,** Eastern Generation Ltd, UK, **J. Cairns, Jr.,** Virginia Polytechnic Institute and State University, USA, **P. A. Chave,** Water Pollution Consultant, UK, **P. Crutzen,** Max-Planck-Institut für Chemie, Germany, **S. J. de Mora,** International Atomic Energy Agency, Monaco, **P. Doyle,** Zeneca Group PLC, UK, **G. Eduljee,** Environmental Resources Management, UK, **M. J. Gittins,** Consultant, UK, **J. E. Harries,** Imperial College of Science, Technology and Medicine, London, UK, **P. K. Hopke,** Clarkson University, USA, **Sir John Houghton,** Meteorological Office, UK, **N. J. King,** Environmental Consultant, UK, **P. Leinster,** Environment Agency, UK, **J. Lester,** Imperial College of Science, Technology and Medicine, UK, **S. Matsui,** Kyoto University, Japan, **D. H. Slater,** Oxera Environmental Ltd, UK, **T. G. Spiro,** Princeton University, USA, **D. Taylor,** Zeneca Group PLC, UK, **T. L. Theis,** Clarkson University, USA, **Sir Frederick Warner,** Consultant, UK.

TITLES IN THE SERIES:

1   Mining and its Environmental Impact
2   Waste Incineration and the Environment
3   Waste Treatment and Disposal
4   Volatile Organic Compounds in the Atmosphere
5   Agricultural Chemicals and the Environment
6   Chlorinated Organic Micropollutants
7   Contaminated Land and its Reclamation
8   Air Quality Management
9   Risk Assessment and Risk Management
10 Air Pollution and Health
11 Environmental Impact of Power Generation
12 Endocrine Disrupting Chemicals
13 Chemistry in the Marine Environment
14 Causes and Environmental Implications of Increased UV-B Radiation

FORTHCOMING:

15 Food Safety and Food Quality

How to obtain future titles on publication

A subscription is available for this series. This will bring delivery of each new volume immediately upon publication and also provide you with online access to each title via the Internet. For further information visit www.rsc.org/issues or write to:

Sales and Customer Care Department, Royal Society of Chemistry, Thomas Graham House, Science Park, Milton Road, Cambridge CB4 0WF, UK

Telephone: +44 (0) 1223 420066
Fax: +44 (0) 1223 423429
Email: sales@rsc.org

ISSUES IN ENVIRONMENTAL SCIENCE
AND TECHNOLOGY

EDITORS: R. E. HESTER AND R. M. HARRISON

14
Causes and Environmental Implications of Increased UV-B Radiation

ROYAL SOCIETY OF CHEMISTRY

ISBN 0-85404-265-2
ISSN 1350-7583

A catalogue record for this book is available from the British Library

© The Royal Society of Chemistry 2000

All rights reserved
Apart from any fair dealing for the purposes of research or private study, or criticism or review as permitted under the terms of the UK Copyright, Designs and Patents Act, 1988, this publication may not be reproduced, stored or transmitted, in any form or by any means, without the prior permission in writing of The Royal Society of Chemistry, or in the case of reprographic reproduction only in accordance with the terms of the licences issued by the Copyright Licensing Agency in the UK, or in accordance with the terms of the licences issued by the appropriate Reproduction Rights Organization outside the UK. Enquiries concerning reproduction outside the terms stated here should be sent to The Royal Society of Chemistry at the address printed on this page.

Published by The Royal Society of Chemistry, Thomas Graham House,
Science Park, Milton Road, Cambridge CB4 0WF, UK
For further information see our web site at www.rsc.org

Typeset in Great Britain by Vision Typesetting, Manchester
Printed and bound by Redwood Books Ltd., Trowbridge, Wiltshire

Preface

The possibility that emissions of chemicals to the atmosphere might cause depletion of ozone in the stratosphere has been recognised since the early 1970s. It was only, however, in the mid-1980s that observations by the British Antarctic Survey showed massive depletion of the ozone column in the austral spring. Since that time there has been a great deal of international activity to quantify better the global distribution of ozone, to monitor and understand loss processes and to devise numerical models capable of predicting future changes in stratospheric ozone and its distribution. These provide an excellent example of atmospheric science in action; the work has involved the three major areas of atmospheric chemistry, *i.e.* laboratory studies which determine the reaction pathways and rate coefficients, field measurements, which define composition and time trends, and numerical modelling which helps to confirm our understanding and provides a predictive capability of future trends. This has been a considerable success for science but much remains to be done. Defining and predicting changes in ozone is, however, in many senses only the beginning of the story. As every secondary school pupil should know, stratospheric ozone serves a crucial role in reducing the intensities of ultraviolet radiation reaching the Earth's surface and therefore protects us and other living organisms from an unpleasant fate. It therefore naturally follows that from our improved understanding of changes in stratospheric ozone, we need to move on and both measure and predict changes in potentially harmful UV-B radiation at the Earth's surface and to examine the effects of such changes on the environment, including living organisms.

This volume of *Issues* charts this research from the changing distribution of atmospheric ozone, to changes in UV-B radiation, and consequent effects on photochemistry and biological systems in the aquatic and terrestrial environments. The first chapter, by John Pyle of the University of Cambridge, summarises current knowledge of stratospheric chemistry relevant to ozone, together with measurements of ozone distribution in the atmosphere. It examines links between climate change and ozone distributions and gives a somewhat pessimistic outlook on the ability of current models to predict trends in ozone concentrations. The second chapter, by Ann Webb of UMIST, examines the processes responsible for the attenuation of UV radiation, the means of measurement of UV-B and current knowledge of trends in UV-B intensities within the atmosphere. The tremendous natural variability in UV-B radiation is pointed out, and clearly

Preface

much remains to be done if we are to predict future trends with confidence.

Chapter 3, by Robert Whitehead of the University of North Carolina and Steve de Mora of the IAEA Marine Laboratory in Monaco, deals with UV photochemistry in the marine environment. This is a highly active research area in itself. It appears that light in the near ultraviolet and short visible wavelengths in the solar spectrum is primarily responsible for most marine photochemical reactions. This observation coupled with known changes in stratospheric ozone has spurred interest in the potential effects of increased UV-B on marine photobiological and photochemical processes. The subsequent chapter by Patrick Neale of the Smithsonian Environmental Research Center and David Kieber of the State University of New York extends this analysis examining the wavelength dependence of photochemical and photobiological processes and showing the development of biological weighting functions which better define the effects of UV-B exposure.

In the subsequent chapter, Jelte Rozema of the Free University of Amsterdam considers the terrestrial environment, examining how enhanced solar UV-B at the Earth's surface may reduce growth of some plants. Results are presented which indicate that greenhouse studies may lead to an overestimation of the effects of increased UV-B, with native plant species in their natural ecosystems showing relatively smaller effects. There are various feedbacks on such processes, such as the decomposition of plant litter and changes to the structure of the vegetation canopy via plant secondary metabolites acting as UV-B receptors. In the final chapter, Brian Diffey of Newcastle General Hospital considers the effects of increased UV-B on skin cancer in human populations. Based on current prediction, incidences of skin cancer will peak around the mid-part of the century with an additional incidence of around 7 per 100 000 population, implying 4200 additional cases of skin cancer per year in the United Kingdom alone.

All of the authors highlight the extreme complexity of this field of science and stress the areas in which knowledge is currently highly incomplete. There are many interactions and feedbacks in the systems that they describe which make prediction of future trends extremely difficult. We have been very fortunate in attracting a number of the leading workers in this field to contribute to this volume of *Issues*. We commend it to our readers as an up-to-date and authoritative summary of the state of the science. It will be of especial value to environmental scientists, policy makers and students seeking knowledge on this subject, which has yet to be assimilated in textbooks in any appreciable way.

Roy M. Harrison
Ronald E. Hester

Contents

Stratospheric Ozone Depletion: a Discussion of Our Present Understanding   1
J. A. Pyle

1	Introduction	1
2	Background	1
3	Detection of Ozone Loss	4
4	Recent Observations of Ozone Loss	7
5	The Future	12
6	Conclusion	15
7	Acknowledgements	16

Ozone Depletion and Changes in Environmental UV-B Radiation   17
Ann R. Webb

1	Introduction	17
2	Historical Interest in UV-B	19
3	Determinants of UV at the Ground	20
4	Changing Factors in Transmission	23
5	UV Radiation at the Ground	26
6	Observations of UV Radiation	27
7	Longer-term Assessments of UV Irradiances	34
8	UV Forecasting	35
9	Conclusion	36
10	Acknowledgements	36

Marine Photochemistry and UV Radiation   37
Robert F. Whitehead and Stephen de Mora

1	Introduction	37
2	Basics of Marine Photochemistry	38
3	Marine Photoreactants, Products and Processes	48

Issues in Environmental Science and Technology No. 14
Causes and Environmental Implications of Increased UV-B Radiation
© The Royal Society of Chemistry, 2000

Contents

4	UV-B Radiation and Global Significance for Marine Biogeochemical Cycles	56
5	Summary and Conclusions	60

Assessing Biological and Chemical Effects of UV in the Marine Environment: Spectral Weighting Functions   **61**
Patrick J. Neale and David J. Kieber

1	Introduction	61
2	Chemical Action Spectra	64
3	Biological Weighting Functions	67
4	Comparative Spectroscopy of Weighting Functions	72
5	Assessment of UV Effects	78
6	Summary and Conclusions	82

Effects of Solar UV-B Radiation on Terrestrial Biota   **85**
Jelte Rozema

1	Evolution of Terrestrial Biota and the Stratospheric Ozone Layer	85
2	Solar UV-B, Polyphenolics, the Pool of Organic Carbon in Terrestrial Environments, and the Balance between Oxygen and Carbon Dioxide in the Earth's Atmosphere	90
3	Current Stratospheric Ozone Depletion: Increased Solar UV-B Radiation Reaching the Earth	90
4	Effects of Enhanced Solar UV-B Radiation on Terrestrial Plants, Adaptations of Terrestrial Plants to Solar UV-B: Evidence from Physiological Studies	92
5	Methodologies for the Study of UV-B Effects on Plants of Terrestrial Biota	95
6	Direct and Indirect UV-B Effects on Terrestrial Ecosystem Processes and Feedbacks, Autotrophic and Heterotrophic Relationships	97
7	Conclusions and Outlook	103
8	Acknowledgements	104

Sunlight, Skin Cancer and Ozone Depletion   **107**
Brian L. Diffey

1	Introduction	107
2	Trends in Atmospheric Ozone and Ambient Ultraviolet Radiation	108
3	Human Exposure to Solar Ultraviolet Radiation	109
4	Effects of Ultraviolet Radiation on Skin	113
5	Risk Analysis of Human Skin Cancer	115

Subject Index   **121**

Editors

Ronald E. Hester, BSc, DSc(London), PhD(Cornell), FRSC, CChem

Ronald E. Hester is Professor of Chemistry in the University of York. He was for short periods a research fellow in Cambridge and an assistant professor at Cornell before being appointed to a lectureship in chemistry in York in 1965. He has been a full professor in York since 1983. His more than 300 publications are mainly in the area of vibrational spectroscopy, latterly focusing on time-resolved studies of photoreaction intermediates and on biomolecular systems in solution. He is active in environmental chemistry and is a founder member and former chairman of the Environment Group of the Royal Society of Chemistry and editor of 'Industry and the Environment in Perspective' (RSC, 1983) and 'Understanding Our Environment' (RSC, 1986). As a member of the Council of the UK Science and Engineering Research Council and several of its sub-committees, panels and boards, he has been heavily involved in national science policy and administration. He was, from 1991–93, a member of the UK Department of the Environment Advisory Committee on Hazardous Substances and is currently a member of the Publications and Information Board of the Royal Society of Chemistry.

Roy M. Harrison, BSc, PhD, DSc (Birmingham), FRSC, CChem, FRMetS, FRSH

Roy M. Harrison is Queen Elizabeth II Birmingham Centenary Professor of Environmental Health in the University of Birmingham. He was previously Lecturer in Environmental Sciences at the University of Lancaster and Reader and Director of the Institute of Aerosol Science at the University of Essex. His more than 300 publications are mainly in the field of environmental chemistry, although his current work includes studies of human health impacts of atmospheric pollutants as well as research into the chemistry of pollution phenomena. He is a past Chairman of the Environment Group of the Royal Society of Chemistry for whom he has edited 'Pollution: Causes, Effects and Control' (RSC, 1983; Third Edition, 1996) and 'Understanding our Environment: An Introduction to Environmental Chemistry and Pollution' (RSC, Third Edition, 1999). He has a close interest in scientific and policy aspects of air pollution, having been Chairman of the Department of Environment Quality of Urban Air Review Group and the DETR Atmospheric Particles Expert Group as well as currently being a member of the DETR Expert Panel on Air Quality Standards and the Department of Health Committee on the Medical Effects of Air Pollutants.

Contributors

S. de Mora, *Marine Environment Laboratory, International Atomic Energy Agency, 4 Quay Antoine 1er, BP 800, MC-98012 Monaco*

B. L. Diffey, *Regional Medical Physics Department, Newcastle General Hospital, Newcastle NE4 6BE, UK*

D. J. Kieber, *Chemistry Department, College of Environmental Science and Forestry, State University of New York, 1 Forestry Drive, Syracuse, NY 13210–2778, USA*

P. J. Neale, *Smithsonian Environmental Research Center, PO Box 28, 647 Contees Wharf Road, Edgewater, MD 21037, USA*

J. A. Pyle, *Centre for Atmospheric Science, Department of Chemistry, University of Cambridge CB2 2EW, UK*

J. Rozema, *Department of Systems Ecology, Institute of Ecological Science, Vrije Universiteit, De Boelelaan 1087, 1081 HV Amsterdam, The Netherlands*

A. R. Webb, *Department of Physics, University of Manchester Institute of Science and Technology, Sackville Street, Manchester M60 1QD, UK*

R. F. Whitehead, *Department of Chemistry and Center for Marine Science, University of North Carolina at Wilmington, Wilmington, NC28403–3297, USA*

Stratospheric Ozone Depletion: a Discussion of Our Present Understanding

J. A. PYLE

1 Introduction

Ozone is an important stratospheric constituent. It absorbs solar radiation strongly at wavelengths around 300 nm, protecting the biosphere from harmful radiation. Ozone is also an important climate gas. The absorption of solar radiation heats the atmosphere and is responsible for the increase of temperature with altitude through the stratosphere. Ozone is also a greenhouse gas, absorbing and emitting in the infrared.

Depletion of ozone was first detected in the Antarctic stratosphere in the mid-1980s. That the depletion is global has since been determined using both satellite and ground-based ozone measurements. This anthropogenic depletion is expected to have important consequences, including leading to enhanced UV at the surface.

In this chapter in Section 2 we will first consider the background to the problem of ozone depletion. The role of ozone is briefly reviewed and the theory of stratospheric ozone depletion is introduced. Ozone loss was first detected in southern polar latitudes. Intensive observational studies then confirmed that the loss was also occurring in middle latitudes and the Arctic. In Section 3 these first observations of ozone depletion in both polar and middle latitudes are discussed, along with the consequent changes in theoretical understanding. More recent measurements of ozone loss up to the end of the 1990s are considered in Section 4. Section 5 speculates about the future state of the ozone layer into the 21st century.

2 Background

The Role of Ozone

Ozone is present in the atmosphere in trace amounts. In the troposphere below around 10 km the ozone mixing ratio is about 50 ppbv (parts per billion by volume). Mixing ratios are much higher in the stratosphere and reach a peak at

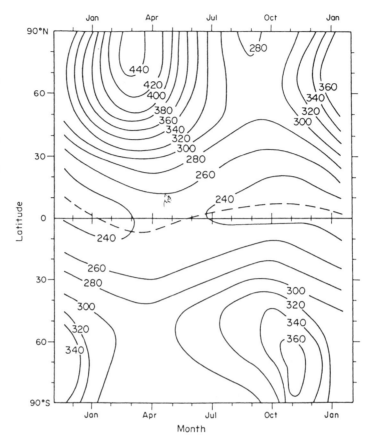

Figure 1 Latitude and seasonal variation in the ozone integrated column amount in Dobson Units. Notice that this figure was produced from observations made prior to the development of the Antarctic ozone hole (see Section 3 and Figure 3, below)

around 10 ppmv (parts per million by volume) in the region known as the ozone layer. The highest concentrations occur in the low stratosphere, between about 15 and 30 km, depending on latitude.

A frequently used measure of ozone is its integrated column amount, the sum of the ozone concentration between the surface and the top of the atmosphere. Values typically range from a little over 200 m atm cm (or Dobson units, DU) in the tropics to greater than 400 DU in the high latitude spring. Figure 1 shows the latitude and seasonal distribution of column ozone. This variation with space and time has been broadly know for about 70 years, following the pioneering measurements of column ozone by Dobson and his co-workers.[1]

The importance of ozone has been recognised for a long time. Ozone plays several important roles. It is toxic and high concentrations at the Earth's surface have implications for the health of both humans and plants. In the stratosphere, ozone absorbs solar ultraviolet radiation strongly, especially at wavelengths less than about 310 nm in the UV-A and UV-B parts of the spectrum. Ozone thus acts as a filter, preventing potentially harmful UV radiation from reaching the surface.

The penetration of radiation around 300 nm depends critically on the ozone

---

[1] G. M. B. Dobson, *Proc. R. Soc. London, Ser. A*, 1930, **129**, 411.

column amount and the precise wavelength of the radiation (since the efficiency of absorption by ozone varies strongly with wavelength around 300 nm). Changes in the ozone column can have a significant impact in changing the penetration of UV to the surface, and for this reason any depletion of the stratosphere ozone column is a cause for concern.

The absorption of solar radiation by ozone also plays a very important role in determining atmospheric structure (ozone is an important climate gas). For example, the rise in temperature with altitude within the stratosphere is a result of the absorption of solar energy by ozone molecules. Absorption of infrared radiation by ozone is particularly important in the lower stratosphere, where changes in ozone are predicted to have a significant impact on surface temperature.[2]

Theory of Stratospheric Ozone

Research in the early 1970s established a good description of the chemical processes responsible for the observed distribution of stratospheric ozone. Before that it had been thought that a sequence of reactions proposed by Chapman[3] could explain the observations of ozone. In Chapman's theory, ozone is produced following the photolysis of molecular oxygen by solar UV radiation at wavelengths less than about 240 nm (equation 1). Two reactions (equations 2 and 3) rapidly interconvert O and O_3 (so that these two species can be thought of together as 'odd oxygen'). Finally, ozone (or 'odd oxygen') is destroyed by the reaction of O and O_3 (equation 4):

$$O_2 + h\nu \rightarrow 2O \qquad \lambda < 242 \text{ nm} \qquad (1)$$

$$O + O_2 + M \rightarrow O_3 + M \qquad (2)$$

$$O_3 + h\nu \rightarrow O_2 + O \qquad \lambda < 1100 \text{ nm} \qquad (3)$$

$$O + O_3 \rightarrow 2O_2 \qquad (4)$$

where M is any third body (usually N_2 or O_2).

During the 1960s it became apparent that these reactions overpredict the observed ozone. The resolution of this discrepancy came through the suggestion that a series of catalytic reactions of the following form could remove ozone:

$$X + O_3 \rightarrow XO + O_2 \qquad (5)$$

$$XO + O \rightarrow X + O_2 \qquad (6)$$

$$\text{net:} \quad O + O_3 \rightarrow 2O_2$$

Thus, these reactions effectively catalyse reaction (4), thereby destroying ozone. The reactive radical species, X, is reformed in the reaction sequence so that the reactions may cycle many times until X is removed by another process. Note that if X is present in the parts per billion range, then one sequence through reactions (5) and (6) will only remove about a part per billion of ozone, very small compared to the abundance of ozone. However, when the cycle operates many thousands of times, it can then exert a controlling influence on stratospheric ozone.

[2] A. A. Lacis, D. J. Wuebbles and J. A. Logan, *J. Geophys. Res.*, 1990, **95**, 9971.
[3] S. Chapman, *Mem. R. Met. Soc.*, 1930, **3**, 103.

X can be, for example, OH,[4] NO[5,6] or Cl.[7,8] These are all reactive radical species which are present in the stratosphere following the breakdown of the source gases H_2O, N_2O and the CFCs (the chlorofluorocarbons).

With the understanding of the role of these catalytic cycles came the realisation that the ozone layer could be perturbed by the introduction of enhanced concentrations of the radical species, X. For example, it was proposed that oxides of nitrogen, emitted directly into the stratosphere by supersonic aircraft, could lead to a large, additional ozone destruction and hence to a reduction in the stratosphere ozone column.[6,9] Similarly, Molina and Rowland[8] showed that the large build-up in the atmosphere of CFCs, then widely used as aerosol propellants, in refrigeration systems and for foam blowing, could also lead to a depletion of the stratospheric ozone layer. They showed that although the CFCs are inert in the troposphere, if they are carried high enough into the stratosphere they can be broken down by ultraviolet radiation to liberate Cl, leading to ozone loss.

During the 1970s, large reductions in ozone were predicted for a proposed fleet of supersonic aircraft or following the large growth in the production of CFCs. A particular concern with the CFCs was that the compounds being produced in largest quantities, $CFCl_3$ and CF_2Cl_2, had atmospheric lifetimes of many decades. Once emitted into the atmosphere, they could therefore remain in the atmosphere for many years as potential ozone-depleters. In the event, only a small fleet of supersonic aircraft was produced, although the topic of ozone depletion by aircraft emissions has re-emerged quite recently as an important issue.[10] However, CFC production and usage continued to grow rapidly throughout the 1970s and the early part of the 1980s. In the next section we present the first evidence that these CFC emissions did indeed have a damaging effect on the stratosphere.

3 Detection of Ozone Loss

The Antarctic 'Ozone Hole'

The first evidence of anthropogenic ozone depletion came dramatically with the discovery by Farman *et al.* of the Antarctic ozone hole.[11] They showed, from observations of column ozone made at Halley Bay (76°S) beginning in 1957, that there had been a rapid decline in the average October column amounts during the late 1970s and early 1980s. Ozone had decreased from values of around 300 DU during the 1960s to about 150 DU in the early 1980s. These measurements created massive scientific interest. The catalytic theories, discussed in Section 2, indicated that the largest ozone depletion should occur in the upper stratosphere.

---

[4] B. G. Hunt, *J. Geophys. Res.*, 1966, **71**, 1385.
[5] P. Crutzen, *Q. J. R. Meteorol. Soc.*, 1970, **96**, 320.
[6] P. Crutzen, *J. Geophys. Res.*, 1971, **76**, 7311.
[7] R. S. Stolarski and R. J. Cicerone, *Can. J. Chem.*, 1974, **52**, 1610.
[8] M. J. Molina and F. S. Rowland, *Nature*, 1974, **249**, 810.
[9] H. S. Johnston, *Science*, 1971, **173**, 517.
[10] IPCC, *Aviation and the Global Atmosphere*, eds. J. E. Penner, D. H. Lister, D. J. Griggs, D. J. Dokken and M. McFarland, Cambridge University Press, Cambridge, 1999.
[11] J. C. Farman, B. G. Gardiner and J. D. Shanklin, *Nature*, 1985, **315**, 207.

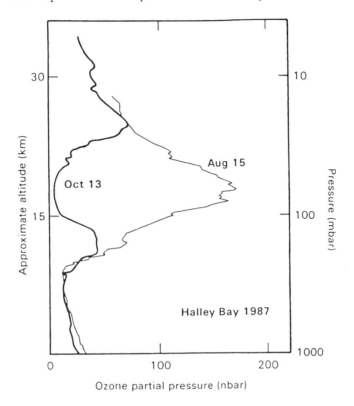

Figure 2 Observations of the vertical ozone profile (in nbar, proportional to the ozone concentration) at Halley Bay (76°S) in 1987. The profile in August is unperturbed; two months later the 'ozone hole' is fully developed, with almost complete depletion between about 16 and 20 km (Taken from SORG[12])

The reported ozone decline was occurring in the lowest part of the stratosphere (see Figure 2)[12] and in springtime, when theory indicated that little loss should occur. A new theoretical understanding was required.

Within two years, the huge research effort had produced a broadly consistent picture of the processes leading to ozone loss (see, for example, the recent review by Solomon[13]). It was shown that the loss of polar ozone was rapid, occurring over just several weeks from late August to mid-October. The loss was largest between about 12 and 20 km. Over part of this altitude range, essentially complete removal of ozone occurred (see Figure 2). Other measurements demonstrated that the loss of ozone was primarily due to a large build-up of active (ozone-destroying) chlorine species in polar latitudes. Elevated ClO concentrations were measured in the Antarctic lower stratosphere from NASA's ER-2 high-flying research aircraft by Anderson and colleagues.[14] Observations showed that the springtime increase in ClO was clearly anticorrelated with the observed decline in ozone. Furthermore, the ClO arose mainly from the breakdown in the stratosphere of the CFCs (the contribution of natural sources to the stratospheric chlorine budget is small, being a little less than 20%). Thus, these measurements confirmed that anthropogenic depletion of ozone was

---

[12] SORG, *Stratospheric Ozone 1990* (United Kingdom Stratospheric Ozone Review Group, second report), HMSO, London, 1988.
[13] S. Solomon, *Rev. Geophys.*, 1999, **37**, 275.
[14] J. G. Anderson, W. H. Brune and M. H. Proffitt, *J. Geophys. Res.*, 1989, **94**, 11465.

occurring and led in 1987 to the adoption of the Montreal Protocol, designed to phase-out production of ozone-depleting substances.

Why the depletion occurred in the springtime Antarctic lower stratosphere was an intriguing question. In particular, why was the concentration of ClO so high when the current theories had suggested that the most abundant forms of chlorine would be HCl and $ClONO_2$, two species which do not destroy ozone? The answer in part arises from the particular meteorological conditions over Antarctica. During the winter, strong westerly winds (the 'polar vortex') circulate around the Antarctic lower stratosphere, isolating the air over Antarctica where the temperatures fall to below 190 K. At these low temperatures, polar stratospheric clouds (PSCs) form, as a co-condensate of water and nitric acid. It was realised that reactions on the surface of the PSCs could turn chlorine into active forms:

$$HCl + ClONO_2 \xrightarrow{PSC} Cl_2 + HNO_3 \qquad (7)$$

The Cl_2 is easily photolysed to liberate chlorine atoms.

It was also appreciated that additional catalytic chlorine cycles, somewhat different to those discussed in the previous section, could be important. In particular, Molina and Molina[15] showed that a cycle involving the chlorine monoxide dimer, Cl_2O_2, was particularly efficient at low temperatures:

$$ClO + ClO + M \rightarrow Cl_2O_2 + M \qquad (8)$$

$$Cl_2O_2 + h\nu \rightarrow Cl + ClO_2 + M \qquad (9)$$

$$ClO_2 + M \rightarrow Cl + O_2 + M \qquad (10)$$

$$2Cl + 2O_3 \rightarrow 2ClO + 2O_2 \qquad (11)$$

$$\text{net:} \quad 2O_3 + h\nu \rightarrow 3O_2 \qquad (12)$$

This, and a cycle involving ClO and BrO, are now believed to explain the majority of the polar loss.

Global Ozone Loss

With the confirmation that the Antarctic ozone loss was indeed due to chemical destruction involving anthropogenic halogen species, attention turned to ozone levels globally. A detailed study of ozone data sets from satellites and from the ground-based network was carried out under the auspices of the World Meteorological Organisation by the International Ozone Trends Panel[16] and by many individual scientists.

Their studies confirmed that ozone depletion was a global phenomenon. Thus, in 1991, the Stratospheric Ozone Review Group (SORG) reported that the analysis of satellite ozone data showed a global loss of 3% between 1979 and 1990 and that, over the same period, the decline in northern mid-latitudes in early

---

[15] L. T. Molina and M. J. Molina, *J. Phys. Chem.*, 1987, **91**, 433.
[16] WMO, *International Ozone Trends Panel 1988* (World Meteorological Organisation Global Ozone Research and Monitoring Project, report 18), WMO, Geneva, 1989.

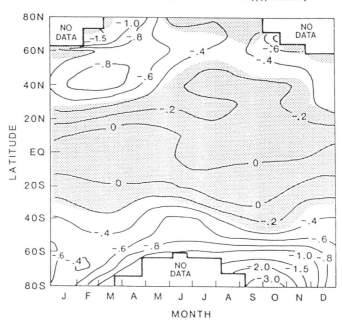

Figure 3 Trend in column ozone by latitude and season in % per decade deduced from the Nimbus-7 TOMS data. The Antarctic loss is a prominent feature but note also the middle latitude losses in both hemispheres and the downward trend at high northern latitudes in winter/spring (Taken from Stolarski et al.[17] with the permission of the AGU)

spring was greater than 8% (these figures can be contrasted with the decline in Antarctic springtime ozone of about 50%). The data also suggested that there had been a decline in ozone over the Arctic. Outside the polar regions, the annual-mean losses of ozone averaged over middle latitudes were comparable in the two hemispheres. No statistically significant loss was detected in the tropics. Figure 3[17] shows the estimated decadal trend in ozone estimated from the Total Ozone Mapping Spectrometer (TOMS) satellite data. The Antarctic decline in springtime is the most obvious feature, but significant losses are seen elsewhere.

4 Recent Observations of Ozone Loss

Following the adoption of the Montreal Protocol in 1987, the Protocol was strengthened during the 1990s in a series of amendments in line with the increasing evidence of the impact on the stratosphere of ozone-depleting substances. Since the beginning of 1996 (1995 in the European Union), production of CFCs and carbon tetrachloride has been phased out in developed countries and schedules for the phase out of replacements are in place. Similarly, production of halons (bromine-containing species used in fire extinguishers, *etc.*) is now phased out and some controls on CH_3Br have been agreed. The impact of this action is that the abundance of chlorine in the stratosphere is now at its peak and should begin to fall slowly (Figure 4). Chlorine in the troposphere is definitely declining.[18,19] The loading of bromine species in the stratosphere is expected to

---

[17] R. S. Stolarski, P. Bloomfield, R. D. McPeters and J. R. Herman, *Geophys. Res. Lett.*, 1015. 1991, **18**.
[18] WMO, *Scientific Assessment of Ozone Depletion: 1998* (World Meteorological Organisation Global Ozone Research and Monitoring Project, report 44), WMO, Geneva, 1999.

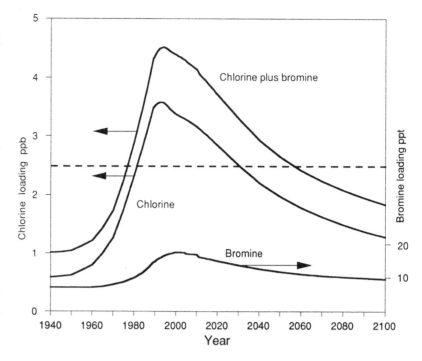

Figure 4 Historical and projected total chlorine and bromine loadings of the troposphere, with projections based on the adoption of the Montreal Protocol and its amendments. Values approximate to the inputs to the stratosphere of reactive chlorine (scaled to the left axis) and reactive bromine (scaled on the right as bromine and on the left as equivalent chlorine). The combined loading is expressed as equivalent chlorine to provide a measure of the combined quantity of ozone-depleting halogens. The line at 2.5 ppbv of equivalent chlorine is indicative of the loading during the latter half of the 1970s before the ozone hole had developed substantially. Note that equivalent chlorine does not fall below this line until the second half of the 21st century (Produced by McCulloch and taken from SORG[20])

peak early in the 21st century.[20] In the context of international regulation of ozone-depleting substances, we report in this section the observations of ozone loss during the 1990s, while in the next section the possible future changes of ozone during the 21st century are considered.

Polar Ozone

The depletion of Antarctic ozone in the spring continues unabated. Every spring, the ozone column inside the polar vortex is depleted and minimum values of around 100 DU are typically reported. Almost complete removal of ozone occurs in the very low stratosphere between about 12 and 20 km. There is some small interannual variability depending on meteorological conditions, so that records in particular measures of ozone depletion (*e.g.* the area covered by the 'hole') are frequently reported. In some ways, these records are misleading. Essentially, the magnitude of the Antarctic ozone depletion should have reached its peak, but will remain close to this peak for decades. While it is difficult to imagine how the polar column loss can get significantly larger (since near complete depletion already occurs in the lower stratosphere), recovery of ozone levels is not expected for many years (see Section 5). The 'ozone hole' is expected to be a feature of observations for at least the first half of the 21st century.

During the 1990s, evidence for chemical depletion of Arctic ozone was

---

[19] S. A. Montzka, J. H. Butler, R. C. Myers, T. M. Thompson, T. H. Swanson, A. D. Clarke, L. T. Lock and J. W. Elkins, *Science*, 1996, **272**, 1318.

[20] SORG, *Stratospheric Ozone* (United Kingdom Stratospheric Ozone Review Group, seventh report), Department of the Environment, Transport and the Regions, London, 1999.

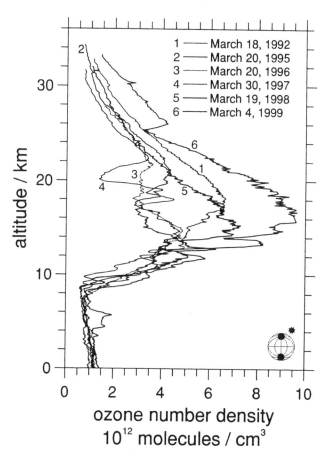

Figure 5 Vertical distribution of ozone measured over Ny Alesund in March for several years in the 1990s. The measurements all correspond to balloon sounding flights made inside the polar vortex (Data courtesy of P. von der Gathen, AWI-Potsdam, and taken from SORG[20])

strengthened and we can now state unequivocally that ozone loss also occurs in the high-latitude late winter and early spring of the northern hemisphere lower stratosphere. The meteorological conditions, a prerequisite for polar loss (see Section 3), are not so severe in the Arctic as in the Antarctic and, for example, the occurrence of PSCs is more sporadic and the polar vortex less strong. Nevertheless, there is now very clear evidence in the Arctic for all the processes which occur in the south: high levels of ClO have been observed[21] and ozone losses of up to 50% in the low stratosphere have been diagnosed.[22] The loss is quite variable from year to year and depends strongly on the particular meteorological conditions. Some very cold winters in the mid-1990s are believed to have led to substantial ozone loss.[22] Figure 5 shows a series of measurements of ozone from Ny Alesund (79°N) during the 1990s. The low ozone in 1994, 1995

[21] J. W. Waters, L. Froidevaux, W. G. Read, G. L. Manney, L. S. Elson, D. A. Flower, R. F. Jarnot and R. S. Harwood, *Nature*, 1993, **362**, 597.
[22] M. P. Chipperfield and J. A. Pyle, *J. Geophys. Res.*, 1998, **103**, 28389.

and 1996 can be attributed to ozone depletion in these years. Note the comparison, and differences, between Figure 5 and Figure 2 showing Antarctic ozone.

The winter of 1999/2000 has again been exceptionally cold in the Arctic lower stratosphere. At the time of writing it is clear that there is potential for very large ozone loss. However, it is the conditions in February and March 2000, when sunlight returns to the most northern latitudes, that will determine the eventual extent of the depletion.

Middle Latitudes

The changes in middle latitude ozone during the 1980s could be reasonably well described by a linear, downward trend.[16] A major perturbation occurred in 1991 when the volcano Mt. Pinatubo erupted, injecting large amounts of SO_2 into the stratosphere. Global ozone was very low in 1992 and, particularly, 1993, attributed to the eruption. The SO_2 was rapidly converted into sulfate aerosol droplets and reactions on the aerosol surface could then enhance reactive chlorine (in a similar fashion to the reaction on PSCs, but with a much reduced efficiency). Model studies have shown that this chemical impact of the eruption can explain the reduced ozone in middle latitudes,[23] and the ozone interannual variability, in the early 1990s. Other studies[24] suggest that the enhanced aerosol may have had a role in changing the atmospheric transport of ozone, thereby contributing to the low ozone observed. It seems likely that both chemical and dynamical processes played a role. The stratospheric aerosol abundance subsequently declined following the eruption and returned to near-background levels by about 1995.

The impact of Mt. Pinatubo was so large that it was impossible to calculate the trend in ozone, arising from halogen compounds, during the period following the eruption. Following reduction in the sulfate aerosol level, new estimates of ozone loss are now being made.[18] Figure 6 shows deviations of the ozone column, averaged over northern middle latitudes, as a function of time since 1979. The very low ozone in 1993, mentioned above, is a prominent feature. The northern hemisphere middle latitude ozone in the late 1990s is about 4–6% lower than around 1980. These values in the late 1990s are a little higher than would be obtained by extrapolating the linear trend determined before the eruption of Mt. Pinatubo, using the data available up to 1991. However, the interpretation of these recent data (for example, as showing a recovery in ozone levels) should be approached with caution (see SORG[20], Section 1.3, for a more detailed discussion). For example, a statistical model is used to determine the trend. Short periods above and below the trend line must be interpreted with great care. A proper test of significance needs to be carried out. It seems unlikely that any reduction in the trend should yet have occurred due to reductions in stratospheric chlorine (which has only just reached its peak, see above). The reduction in the

[23] S. Solomon, R. W. Portman, R. R. Garcia, L. W. Thomason, L. R. Poole and M. P. McCormick, *J. Geophys. Res.*, 1996, **101**, 6713.

[24] P. Hadjinicolaou, J. A. Pyle, M. P. Chipperfield and J. A. Kettleborough, *Geophys. Res. Lett.*, 1997, **24**, 2993.

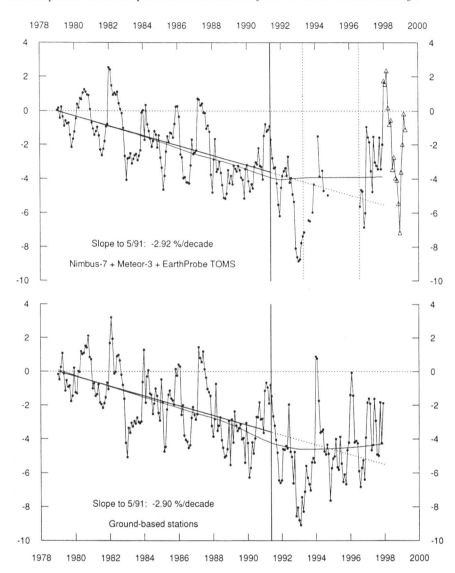

Figure 6 Deviations in the total integrated ozone column, area weighted between 25°N and 60°N. The upper curve is based on satellite data and the lower curve on data from ground-based stations. The seasonal trend model used to fit the ozone data included allowance for the effects of the solar cycle and the quasi-biennial oscillation. The solid straight line represents the least squares fit to the deviations up to May 1991 and is extended as a dotted line through to December 1997. The thinner smoother line is from a lowess local regression fit and does not include the TOMS data after January 1998 (triangles) (Updated from WMO[18] courtesy of L. Bishop, Allied Signal, and from SORG[20])

growth of stratospheric chlorine, consequent on the Montreal Protocol, should be secondary to the chemical impact of reduced sulfate aerosol in the late 1990s. Changes in atmospheric circulation, not adequately included in the statistical model, may also be important. SORG states[20] that 'for all these reasons, we do not think that the observational record can be interpreted as showing any evidence of an ozone recovery in response to reductions of ozone-depleting substances'.

Nevertheless, reductions in the atmospheric abundance of ozone-depleting substances are definitely occurring and it is therefore reasonable to expect to see an impact on stratospheric ozone. In the next section we consider the possible future state of the ozone layer during the first half of the 21st century.

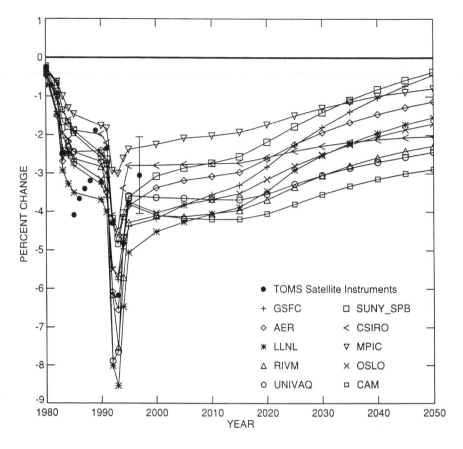

Figure 7 Percentage change in 65°S–65°N annual average column ozone relative to the year 1979, for a range of 2D models for the period 1980 to 2050. The TOMS observations for the period 1979 to 1997 are indicated by the solid circles (Taken from WMO[18])

5 The Future

The vast majority of anthropogenic ozone-depleting substances are now controlled under the Montreal Protocol. Emissions of many compounds should now be essentially zero and the stratospheric halogen burden has reached its peak. As it falls, ozone levels should recover. In this section, we consider the question of recovery: how quickly will it become evident and what are the uncertainties?

In the recent WMO assessment of stratospheric ozone,[18] a number of 2D atmospheric models were used to study ozone recovery. Given a scenario for the future halogen emissions, the models calculated global ozone levels out to the middle of the 21st century. Figure 7 shows some results from the models. All the models indicate that the global minimum in ozone should already have occurred, after the eruption of Mt. Pinatubo. Ozone levels in all the models then gradually increase. The increase is, however, very slow and by 2050 none of the models has recovered to pre-1980 values. The slow recovery is, of course, related to the long residence time in the stratosphere of the ozone-depleting halogen species (see Figure 4). The uncertainty in the calculation is demonstrated by the wide range of model results. Nevertheless, ozone levels below those present in the 1970s are predicted for the next several decades in all the models. The models also considered the impact of future volcanic eruptions. They found that an eruption similar to Mt. Pinatubo could again cause large ozone depletion, especially if it was to occur in the next 20 or so years before halogen levels have dropped significantly.[18]

Notice that the models do not reproduce the details of the TOMS ozone observations in the late 1980s particularly well. Unless models can fully explain the past changes in ozone, their ability to predict the future must be questioned. Indeed, the detailed mechanisms responsible for the middle latitude decline are still controversial and must be understood in order to validate our predictive capability. At issue is the extent of *in situ* middle latitude ozone depletion *versus* the role of polar processes on middle latitudes. Furthermore, the contribution of dynamical changes is also incompletely understood. In a 2D modelling study, Solomon *et al.*[25] have shown that nearly all the observed changes in ozone at northern mid-latitudes during the 1980s and 1990s can be explained quantitatively by the slowly changing chlorine and bromine loading of the stratosphere, modulated by the more rapid changes in surface area of the stratospheric sulfate aerosol. Other studies have investigated dynamical mechanisms. For example, Hood *et al.*[26] have shown that changes in the wintertime circulation could explain up to half of the northern mid-latitude ozone trends. Long-term changes in tropopause height[27] and the strength of the North Atlantic oscillation[28] have also been suggested as being involved in the observed ozone trend. Using a 3D atmospheric model, Hadjinicolaou *et al.*[24] showed that changes in transport did influence middle latitude ozone during the 1990s. For example, their model calculation reproduced the very low ozone of 1993 purely by a transport mechanism. 'Dilution' (mixing of air from polar into middle latitudes) from the ozone-depleted polar vortex will also be important and its contribution will vary interannually.[29]

It seems clear that a variety of chemical and physical mechanisms are involved in the middle latitude ozone decline. Until a full, quantitative understanding of the importance of all the processes is established, we will not be able to model the state of the global ozone layer in future decades with confidence.

Increasingly sophisticated models are being used to study stratospheric ozone and several coupled 3D circulation models have recently addressed the future state of polar ozone loss.[18] These calculations, starting with the calculation of Shindell *et al.*,[30] have highlighted the importance of the chemistry/climate feedback. Observations of atmospheric temperature have now confirmed that there has been a downward trend in temperatures in the lower stratosphere of about 0.6 K/decade since 1980.[18] This is partly due to the reduction in ozone in

---

[25] S. Solomon, R. W. Portmann, R. R. Garcia, W. Randel, F. Wu, R. Nagatani, J. Gleason, L. Thomason, L. R. Poole and M. P. McCormick, *Geophys. Res. Lett.*, 1998, **25**, 1871.

[26] L. L. Hood, J. P. McCormack and K. Labitzke, *J. Geophys. Res.*, 1997, **102**, 13079.

[27] W. Steinbrecht, H. Claude, U. Köhler and K. P. Hoinka, *J. Geophys. Res.*, 1998, **103**, 19183.

[28] C. A. Appenzeller, A. K. Weiss and J. Staehelin, *Geophys. Res. Lett.*, 2000, in press.

[29] B. M. Knudsen, I. S. Mikkelsen, J.-J. Morcrette, G. O. Braathen, G. Hansen, H. Fast, H. Gernandt, H. Kanzawa, H. Nakane, E. Kyro, V. Dorokhov, V. Yushkov, R. J. Shearman and M. Gil, *Geophys. Res. Lett.*, 1998, **25**, 4501.

[30] D. T, Shindell, D. Rind and P. Lonergan, *Nature*, 1998, **392**, 589.

Figure 8 Minimum total column ozone south of 65°S for a range of 3D models. The GISS model results are shown as open circles. The bars at 1990 and 2015 show the one standard deviation variability of the ECHAM/CHEM model over the last 10 years of their simulations. The CNRS model results are shown by the large open circle, with the bars indicating the range of modelled values over a five-year simulation. The UK Meteorological Office (UKMO) results are indicated by the triangles. The TOMS observations for 1979–1997 are shown by the filled circles [Based on results presented in Chapter 12 of WMO[18] (from where further details can be obtained) and modified in SORG[20]]

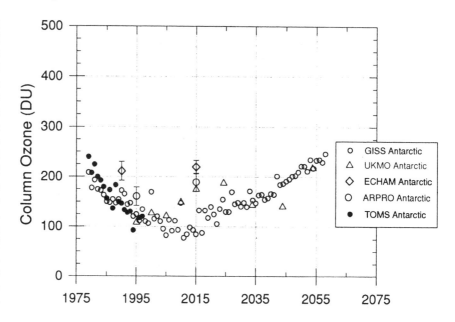

that region; increases in greenhouse gases may also have contributed. Polar ozone loss, as discussed in Chapter 3, is favoured by low temperatures and a strong polar vortex. If the occurrence of these meteorological conditions increases in the future then this could, to a greater or lesser extent, compete with any reduction in stratospheric halogens. In the Shindell et al.[30] study, the northern hemisphere polar vortex was predicted to be colder and more stable during the next few decades. In consequence, maximum ozone depletion was predicted to occur not exactly in coincidence with maximum halogen loading (i.e. around 2000) but more than a decade later, approaching 2020. The model used by Shindell et al.[30] was simplified in some respects. Other 3D chemistry/climate models with more detailed descriptions of stratospheric chemistry have repeated their calculations and found broadly the same results, as shown in Figure 8. Again, there is a model uncertainty associated with these predictions.

It is clear that recovery of ozone cannot be thought of simply in terms of the return of stratospheric halogen levels to, for example, the levels before Antarctic ozone depletion became evident. The atmosphere of the 21st century will differ in important respects from the atmosphere of the middle to late 20th century. The growth of greenhouse gas concentrations will change the composition of the stratosphere and, most importantly, stratospheric temperatures. These factors will influence ozone concentrations and ozone loss processes. Thus, we cannot expect a simple, monotonic return of ozone to previous values as the halogen loading declines in response to the Montreal Protocol. Indeed, calculations suggest that the recovery to previous ozone levels will be delayed beyond that expected from halogen levels alone.

A particularly interesting example of coupled chemistry/climate change has been suggested by Waibel et al.[31] They have pointed out that one consequence of a colder polar lower stratosphere could be to make the Arctic behave more like the Antarctic. In particular, denitrification could become a more regular feature of the Arctic polar vortex. Denitrification occurs in the Antarctic winter when nitrogen species, in the form of nitric acid, are removed from the gas phase into particles (initially the PSCs and then, as temperatures drop, ice particles) which then sediment out of the stratosphere. This removal of nitrogen species has an important chemical consequence in that high levels of active chlorine can be maintained for a longer period, thereby sustaining ozone loss. When nitrogen oxides are present, the ozone-destroying ClO can be removed into $ClONO_2$, thereby halting the ozone loss. Thus, not withstanding a reduction in stratospheric halogens, a future decrease in temperature leading to denitrification could actually enhance Arctic loss in the coming decades.

Nevertheless, early detection of ozone recovery is an important aim during the next decades. The first effects of the Montreal Protocol are evident in the deceleration of the growth of ozone-depleting substances and now in the decline of their atmospheric concentrations. The next, important step will be the detection in the stratosphere of the impact of this halogen decline.

6 Conclusion

In this article, the behaviour of the stratospheric ozone layer during the last few decades, and the present understanding of that behaviour, has been reviewed. The discovery of the Antarctic ozone hole in 1985 was a major demonstration of anthropogenic global change. It is now known unequivocally that the ozone loss is driven by chemical reactions involving the breakdown products of the CFCs, and other halogen carriers, which have been introduced into the atmosphere during the last few decades. We now also know that these same compounds are involved in a global decline in stratospheric ozone amounts. In response to the improved scientific understanding, the Montreal Protocol and its amendments have now regulated the major ozone-depleting substances and in many cases their atmospheric abundances are now declining. However, the long atmospheric lifetime of the CFCs means that recovery will be very slow, taking many decades. We expect that large seasonal loss of ozone in Antarctic will continue until at least the middle of the next century. Large ozone losses can also be expected in the Arctic springtime, especially when temperatures are low and favour the production of PSCs. Annually averaged ozone in middle latitudes is now about 4–6% less than the values around 1980. There seems little prospect of a rapid recovery; middle latitude ozone is expected to be depressed for at least several more decades.

The details of the recovery of the ozone layer remain uncertain, not least because we cannot fully explain the observations of ozone in the recent past. Two things are clear: recovery will be slow, and it may well be influenced by other stratospheric changes in consequence of the growth of greenhouse gases.

[31] A. E. Waibel, T. H. Peter, K. S. Carslaw, H. Oelhaf, G. Wetzel, P. J. Crutzen, U. Poschl, A. Tsias and E. Reimer, *Science*, 1999, **283**, 2064.

7 Acknowledgements

The author thanks the Department of the Environment, Transport and the Regions (DETR) for support. This report relies heavily on the report recently prepared for the DETR by the Stratospheric Ozone Review Group (SORG), which the author chairs. Several figures are reproduced from that and an earlier SORG report. Crown copyright is reproduced with the permission of the Controller of Her Majesty's Stationery Office.

Ozone Depletion and Changes in Environmental UV-B Radiation

ANN R. WEBB

1 Introduction

Ultraviolet (UV) radiation, that region of the electromagnetic spectrum between visible light and X-rays, is a natural component of sunlight and thus a part of the environment in which life evolved and to which it is adapted. Extraterrestrial solar radiation contains radiation of all UV wavelengths, but the solar spectrum is modified as it passes through the Earth's atmosphere so that the shortest wavelengths of solar radiation to penetrate as far as the lower troposphere (the bottom few kilometers of the atmosphere, where life exists) are in the UV-B portion of the spectrum (Figure 1). Ultraviolet-B radiation is defined by the Commission Internationale d'Eclairage (CIE) as the waveband 280–315 nm, while 315–400 nm is classed as UV-A radiation and shorter wavelengths 200–280 nm are designated UV-C. However, the solar spectrum is a continuum across the UV spectral region, and the influences on or responses to UV radiation are functions of wavelength that often cross the UV-B–UV-A boundary defined above, so using the strict CIE definitions to discuss solar UV causes and effects would be either misleading or require constant clarification. For the purposes of this article the UV-B–UV-A boundary will be taken as a somewhat ill-defined region 'round about 320 nm' where the solar spectrum approaches a plateau and many action spectra for UV effects tend to zero or immeasurable levels of response (Figure 1); thus our UV-B radiation certainly incorporates but is not exclusive to the CIE waveband.

The major atmospheric absorber determining the shape of the UV part of the ground-level solar spectrum is ozone (O_3), which is found in the lowest 50 km of the atmosphere (troposphere and stratosphere), but predominantly in the stratosphere where about 90% of the ozone in a vertical column resides. The greatest concentration of the gas is in the layer between 15–30 km altitude. All the shortest wavelengths of solar radiation are absorbed in the outermost layers of

---

Issues in Environmental Science and Technology No. 14
Causes and Environmental Implications of Increased UV-B Radiation
© The Royal Society of Chemistry, 2000

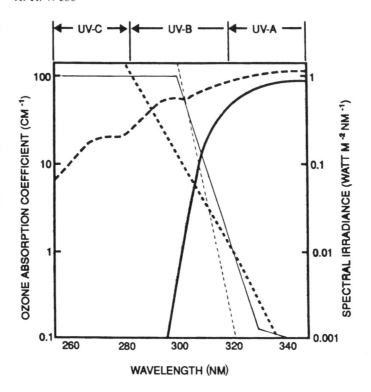

Figure 1 Schematic representation of the extraterrestrial (thick dashed) and ground-level (thick solid) solar spectra and the ozone absorption spectrum (thick dotted). Biological action spectra for erythema (thin solid) and DNA damage (thin dashed) are shown as percent relative response, using the left vertical axis

the atmosphere, mainly by oxygen (N, N_2 and N_2O also absorb the shortest wavelengths at altitudes above 100 km), leaving a spectrum at the top of the stratosphere which is still rich in UV-C and UV-B radiation ($\lambda > 200$ nm). Below 60 km, oxygen still absorbs in the Herzberg continuum (200–242 nm), resulting in two ground-state oxygen atoms which are important for the formation of ozone, but it is the ozone itself which is the major absorber, with three absorption bands, two in the UV and the Chappuis band in the visible region. Ozone absorption is strongest in the Hartley band, a continuum of decreasing absorption coefficient with increasing wavelength from 200 to 310 nm. This merges into the line spectrum of the Huggins band (310–400 nm). In combination the two bands prevent radiation of $\lambda < 280$ nm from reaching the Earth, but have little effect on radiation of $\lambda > 340$ nm. The steep reduction in ozone absorption coefficient is mirrored by the steep increase in radiation reaching the ground across the UV-B waveband (Figure 1). The basic laws of physics then determine that in an otherwise static atmosphere a decrease (or increase) in ozone in the atmosphere would lead to an increase (or decrease) in UV-B radiation reaching the ground. Since the high energy of UV-B photons enables them to drive many chemical and biological reactions, a significant increase in their arrival in the biosphere could affect a number of aspects of the environment, often in a detrimental way. The potential for, and then observations of, a decreasing stratospheric ozone layer have therefore been cause for concern for the past two decades.

2 Historical Interest in UV-B

The ultraviolet portion of the solar spectrum accounts for less than 10% of the total solar energy incident at the Earth's surface, and despite the high energy of the individual photons the UV-B spectral region contains less than 1% of the solar energy. In terms of providing energy for heating and driving atmospheric motion it is therefore insignificant and largely ignored by the meteorologist as a feature of weather or climate. The interest in, and attempts to measure, UV-B in the past have been driven either by the medical and environmental communities as part of UV effects research, or within the atmospheric community as a means and then as a by-product of ozone measurements.

In the first half of the 20th century, UV radiation was first recognized for its ability to cure rickets, then for its detrimental contribution towards the development of skin cancer, and it was a group of physicians who, in the 1920s, first made and reported on UV measurements at different sites,[1] using an acetone solution of methylene blue as a kind of dosimeter. At about the same time an instrument to measure stratospheric ozone was being developed within the atmospheric community. The resulting Dobson spectrophotometer[2] is still used as the basis for the ground-based ozone monitoring network, and uses relative measurements of radiation at pairs of UV wavelengths to derive the column ozone. This instrument does not provide absolute measurements of UV, but in recent years the Brewer spectrophotometer[3] has found increasingly widespread use as an alternative to the Dobson instrument. The Brewer works on the same basic principle as the Dobson, but also has the capability to make absolute measurements of the UV spectrum in the wavelength range from 290 to 325 (or 365) nm. Thus an increasing number of ozone monitoring sites now also measure UV-B radiation.

The International Geophysical Year (1957) saw the establishment of a worldwide ozone monitoring network (using the Dobson spectrophotometer), providing for the understanding of ozone climatology against which current changes are judged. Shortly afterwards, dedicated spectral UV measurements were made at Davos in Switzerland over a period of several years, providing the understanding of factors affecting UV radiation at the surface that has been the basis of UV ozone work ever since.[1] At the time the comprehensive Davos measurements seemed to give a sufficient understanding of solar UV-B radiation and the programme ceased. It took medical concerns to revive the interest in measuring solar UV-B during the 1970s.

Reports of an action spectrum for erythema peaking in the UV-B, and a latitudinal gradient in skin cancer incidence, coincided with the first doubts for the safety of the stratospheric ozone layer. The initial threat was thought to come from the then proposed fleets of supersonic aircraft, whose major exhaust products (NO_x) were cited as potential destroyers of ozone in the catalytic cycle:

---

[1] A. R. Webb, in *Measurements and Trends of Terrestrial UV-B Radiation in Europe*, ed. B. L. Diffey, OEMF, Milan, 1996, pp. 9–20.
[2] G. M. B. Dobson, *Proc. Phys. Soc.*, 1931, **XLIII**, 324.
[3] A. W. Brewer, *Pure Appl. Geophys.*, 1973, **106–108**, 919.

$$NO + O_3 \rightarrow NO_2 + O_2$$
$$NO_2 + O \rightarrow NO + O_2$$

net: $O + O_3 \rightarrow 2O_2$

Mechanisms of ozone loss are described in more detail in Chapter 1. In the event, few supersonic aircraft were built and operated, but the potential of increased UV-B combined with the realization of its harmful health effects prompted new efforts to measure UV. The enduring result of those efforts is the broadband radiometer for measuring erythemally effective radiation: initially the Robertson-Berger meter, but now available in several makes and models.[4]

Erythemal UV measurement networks were set up in Australia and USA in the late 1970s, to determine UV climatologies (the networks were not intended to identify trends in UV radiation: the threat of large fleets of supersonic aircraft had subsided and the ozone layer did not appear to be in any danger; increased skin cancer incidence was a danger to immigrant populations, *e.g.* in Australia, and a result of increasing leisure time and foreign vacations for other fair-skinned populations). Thus, after six years of measurements to establish climatology the Australian network ceased to operate in 1981, while the US network ran from 1974 to 1985. Elsewhere, UV dosimeters were developed and used in epidemiological studies, and solar UV measurements were made at a few discrete sites in individual studies, but it took the shock of massive ozone depletion to stimulate renewed interest in solar UV-B radiation, its incidence at the Earth's surface, and the subsequent effects it could cause.

In 1985 a paper was published in the journal *Nature* describing the observation of large ozone losses over the Antarctic during the springtime,[5] a phenomenon subsequently dubbed the 'ozone hole'. The ozone records from scientific bases in the Antarctic showed that the sudden spring ozone loss could be identified as far back as the late 1970s, and had been getting progressively worse. Since that first clear identification of ozone loss the ozone hole has continued to get bigger (greater areal extent), deeper (more loss in a column, *i.e.* lower measured ozone values), and lasts longer, with ozone remaining depleted into the summer months. Ozone loss has also been observed at mid-latitudes and in the Arctic, although not to the same extent as over Antarctica (see Chapter 1). A great deal of research has been invested in explaining the ozone losses since 1985, coupled with political action to prevent ever-increasing depletion,[6] and the resulting increased UV-B at the Earth's surface. However, observing and quantifying solar UV-B radiation has never been a routine part of atmospheric monitoring, and the vast majority of measurement sites in operation today were established post-1985 in response to the ozone scare. Thus data records are short on the time scales it takes to establish climatology, and then identify deviation from the climatological norm.

3 Determinants of UV at the Ground

The UV radiation incident at the Earth's surface is determined by its angle of incidence (the solar zenith angle), and the same processes of radiative transfer that

---

[4] A. R. Webb, in *Advances in Bioclimatology* 5, ed. A. Auliciems, Springer, Berlin, 1998, pp. 7–60.
[5] J. C. Farman, B. G. Gardiner and J. D. Shanklin, *Nature*, 1985, **315**, 207.
[6] Montreal Protocol on Substances that Deplete the Ozone Layer, 1987.

affect every other waveband in the solar spectrum traversing the Earth's atmosphere: absorption and scattering. The solar zenith angle (SZA) is the angle between the normal to the surface (zenith position) and the position of the sun in the sky. When the sun is directly overhead (at noon at the equator during the equinox), the SZA is zero and the direct solar beam strikes the surface at right angles with all the energy concentrated on the smallest possible area. As the SZA increases the same energy is spread over ever increasing areas of the surface and the radiative energy per unit area at a site decreases. Thus latitude, season and time of day, which control SZA, are major determinants of the UV at the surface and dictate the global pattern of UV and the underlying climatological cycles upon which all other factors act. In most regions of the world, changing a latitude band will result in a greater change in exposure to UV radiation than local ozone depletion; for example, moving from Aberdeen to London for a summer break is the UV equivalent of a sudden 10% ozone depletion in Aberdeen.[7]

The SZA also determines the pathlength of radiation through the Earth's atmosphere: the larger the SZA, the longer the slanting path that the radiation travels through a medium (the atmosphere), and the more absorption and scattering takes place. Even in a pristine atmosphere, scattering will take place from the air molecules (so called Rayleigh scattering) in a manner that is dependent on (wavelength)$^{-4}$, so Rayleigh scattering is much stronger at UV wavelengths than at longer wavelengths. In the visible waveband, blue light is scattered more than red, giving the blue sky appearance. During Rayleigh scattering, approximately half the radiation is scattered in the forward direction, reaching the surface as diffuse radiation, and the rest is back scattered to space. At large SZA, virtually all the UV-B radiation is removed from the direct solar beam by scattering, and the small amount that does reach the surface is diffuse. Even when the SZA is small the wavelength dependence of Rayleigh scattering means that over half the total surface UV-B radiation is diffuse, the direct beam accounting for the remaining, smaller portion. The wavelength-dependent scattering becomes less dominant as the scattering particles become larger, and Mie scattering (which depends on the ratio of wavelength to particle size, and the number of scattering particles in the atmosphere) takes over. This effect is more local and variable as it depends on the aerosol distribution in the atmosphere, which changes with time and place. Mie scattering projects the majority of radiation in a forward direction (towards the Earth's surface), although by a longer route than the direct solar beam. There is little wavelength dependence to the process; thus clouds look white.

Reflection at the Earth's surface returns UV radiation to the atmosphere, where some of it undergoes further scattering to return once more to the surface, enhancing the incoming flux. Reflectivities of many surfaces are lower in the UV than at longer wavelengths[8,9] and surface reflectivity has little influence on the UV reaching the surface. However, snow has a high albedo[10] (about 40–100% depending on the type and age of snow) and snow-covered surfaces can

---

[7] Stratospheric Ozone Review Group, *Stratospheric Ozone 1996*, The Stationery Office, London, 1996.
[8] A. R. Webb, I. M. Stromberg, H. Li and L. M. Bartlett, *J. Geophys. Res.*, 2000, **105**, 4945.
[9] U. Feister and R. Grewe, *Photochem. Photobiol.*, 1995, **62**, 736.
[10] M. Blumthaler and W. Ambach, *Photochem. Photobiol.*, 1988, **48**, 85.

significantly enhance the UV reaching the ground. Where the local climate provides for snow and non-snow covered seasons, the changing albedo becomes a factor in the local UV climatology. Reflections can cause an average increase of 20–50% at some sites.[11]

Another local and variable influence is cloud. The effects of cloud are easily observed in the visible waveband, and the effects in the UV are very similar. Cloud can both attenuate and enhance the radiation reaching the surface. Enhancement is usually brief and occurs when clouds (usually cumulus) close to the direct solar beam increase forward scattering (from the cloud sides). The cloud then frequently passes in front of the sun, blocking the direct solar beam and the incident radiation is reduced for the period of the cloud's passage before the sun. The more prominent cloud effect is a reduction in surface radiation. A cloud layer will backscatter (reflect) radiation to space, and will scatter and absorb radiation within the cloud, the degree to which these processes occur depending on the thickness and microphysical properties of the cloud, and the wavelength of radiation under consideration. As cloud can change greatly from moment to moment, and season to season while still being within the climatological norm (at least at mid-latitudes), it introduces tremendous variability into the time sequence of radiation reaching the surface.

Then the wavelength-specific process of absorption must be considered, with ozone the major absorber in the UV waveband (Figure 1). If the ozone layer was completely static it would only cause changes in the solar spectrum through interaction with the SZA (increased pathlength increasing absorption). However, this is not the case and the stratospheric ozone layer has its own natural patterns of global distribution, and its own natural cycles and variabilities. Ozone varies with the 22-year sunspot cycle, with the quasi-biennial oscillation (QBO), and on a seasonal basis (for example, in northern mid-latitudes the ozone maximum occurs in spring and the minimum in autumn). The natural seasonal cycle and global distribution of ozone (see Chapter 1) results from a balance between photochemical production and destruction, and transport. Ozone production requires solar radiation of wavelengths < 240 nm to break the molecular oxygen bond:

$$O_2 + h\nu \rightarrow 2O \qquad (1)$$

$$O + O_2 + M \rightarrow O_3 + M \qquad (2)$$

thus the main source region for ozone is at low latitudes in the middle and upper stratosphere (maximum production is at an altitude of about 40 km). The large-scale stratospheric circulation transports much of this ozone to its sink regions at higher latitudes, and lower altitudes, where ozone destruction is also a photolysis event:

$$O_3 + h\nu \rightarrow O + O_2 \qquad (3)$$

but is driven by longer wavelength radiation (in the UV and visible wavebands). This is the reaction that wholly or (as wavelength increases) partially removes UV radiation from the solar spectrum, protecting the Earth from damaging

---

[11] H. Schwander, B. Mayer, A. Ruggaber, A. Albold, G. Seckmeyer and P. Koepke, *Appl. Optics*, 1999, **38**, 3869.

radiation. The repetition of steps (2) and (3) leads to heating of the stratosphere with no net loss of ozone, so the ozone cycle is also a major determinant of the temperature structure of this part of the atmosphere. Further ozone loss comes from the reaction of atomic oxygen with ozone:

$$O + O_3 \rightarrow 2O_2 \tag{4}$$

Variability (or long-term changes) in stratospheric circulation can cause variability (or potential changes) in global column ozone amounts as it affects the dynamic balance between photochemistry and transport. Tropospheric transport (weather systems) also cause large day-to-day variations in ozone measured at a single site; for example, low ozone is associated with high pressure systems in the troposphere. The ozone layer is not, therefore, a constant either spatially or on short time scales, and any long-term changes or trends have to be identified and judged against the range of natural variabilities which occur on many timescales and could act together or in opposition at different times.

The ozone cycles and fluctuations, added to the SZA cycles and cloud and aerosol fluctuations, lead to a complex mixture of factors which determine UV at the surface, and a very wide range of irradiances that might be classed as 'within normal expectation' for a given time and place.

4 Changing Factors in Transmission

Stratospheric Ozone Depletion

It is against this background of complex interaction between multiple factors affecting UV at the surface that ozone depletion must be viewed. When, in 1985, the Antarctic springtime ozone loss was undeniably illustrated, even against the expected year-to-year variations in ozone, there was no immediate explanation for the observations, and great concern that similar ozone depletion might occur elsewhere in the stratosphere. It had been proposed in 1974[12] that reactions involving degradation products of chlorofluorocarbons (CFCs) could destroy ozone in the stratosphere—work that lead to the Nobel Prize for the authors in 1995—but when ozone loss was first identified (in the lower stratosphere), photochemical theory was predicting loss in the upper stratosphere through reactions involving atomic oxygen and the chlorine monoxide radical, ClO. Today it is widely accepted that ClO does play an important role in the catalytic cycles which deplete ozone. However, the reactions are not the same as those initially proposed, and do not only involve atomic oxygen. An important catalyser for large ozone loss in polar regions is the presence of polar stratospheric clouds (PSCs), which provide sites on which fast heterogenous reactions can occur and rapidly increase the ozone-destroying molecules in the stratosphere. The current understanding of ozone depletion processes, and the state of the ozone layer, have been described in the chapter by Pyle.

It is clear that, to date, fears of widespread, massive ozone depletion have not been fulfilled. Nonetheless, the Antarctic ozone hole continues to appear every year, a less extreme depletion occurs in the northern polar regions, and ozone

---

[12] M. Molina and F. S. Rowland, *Nature*, 1974, **249**, 810.

depletion is observed at all other latitudes outside the Tropics. The prognosis for the recovery of the ozone hole, based on the reduction of ozone depleting chemicals in the atmosphere, is for a return to normal ozone levels by about the year 2050 (see Chapter 1). However, there are other facets of the climate and atmospheric constituents which are, or could, also change and interact with the ozone cycle to alter this prediction.

Aircraft Emissions

Although the fleets of supersonic aircraft proposed in the 1970s did not materialize, there has been a vast increase in air traffic since that time, and there have been more recent proposals for increased numbers of supersonic aircraft flying at altitudes of 15–21 km in the stratosphere (subsonic aircraft generally fly in the troposphere). Assessments of the impact of increased air traffic[13] have shown that the exhaust emissions have the potential to alter atmospheric composition, which could then influence ozone (and climate). Increases in the atmospheric content of water, NO_x, sulfur gases, aerosols and soot could affect the heterogenous and gas phase chemistry that helps determine ozone concentrations in the lower stratosphere and upper troposphere, which in turn would influence the temperature of these regions and hence the transport within them. Emissions occur where the aircraft fly, and are thus concentrated in northern hemisphere mid-latitudes, but atmospheric transport can spread the pollutants more widely, particularly in the stratosphere. The models used to estimate the impact of assumed future fleets of supersonic aircraft, and increased subsonic aircraft, showed that the impact on ozone was dependent on the altitude of flight. Emissions in the troposphere (subsonic) increase tropospheric ozone at the cruise altitude, a response that is almost linear with emissions of NO_x. The maximum yearly average increase of this effect was estimated to be 13% by 2050, or an increase in column ozone of +1.2%. In the stratosphere the projected fleet of 500 supersonic aircraft were estimated to give a slightly negative (less than 1%) change in column ozone by the year 2015, although this was the result of more negative change at about 30 km combined with an increase in ozone lower in the stratosphere at 20 km. Water was the most important emission associated with ozone loss, NO_x providing a smaller contribution at these higher altitudes. Altogether the modelled response of ozone to aircraft emissions is small when compared to the depletion due to halogens.

Interactions with Greenhouse Warming

The interactions between stratospheric ozone loss and other observed or potential atmospheric changes associated with global warming are complex and uncertain. The increasing carbon dioxide that acts to warm the surface in the troposphere has the opposite effect on the stratosphere and leads to cooling. Ozone loss in itself also cools the stratosphere since less radiation (of both solar

---

[13] Intergovernmental Panel on Climate Change, *Special Report on Aviation and the Global Atmosphere*, ed. J. E. Penner, D. H. Lister, D. J. Griggs, D. J. Dokken and M. McFarland, Cambridge University Press, 1999, pp. 373.

and terrestrial wavelengths) is absorbed. A cooler stratosphere means that, at least at high latitudes, there is an increased chance of PSCs forming and existing for more prolonged periods than before.[14] This would support ozone loss processes, accelerating ozone depletion and cooling the stratosphere even further in a positive feedback effect. In addition to reduced absorption of radiation, depleted ozone also means reduced emission of radiation; thus ozone depletion acts to increase the UV radiation reaching the lower atmosphere, but to decrease the infrared radiation. The net result of ozone loss is to cool the climate, and ozone loss is calculated to have offset 15–30% of the warming due to greenhouse gases since the late 1970s (the uncertainties come from different types of models with varying complexity, and uncertainties in the model inputs, *e.g.* the amount and vertical profile of the ozone losses).[15]

In addition to its interaction with ozone in the stratosphere, global warming has other potential influences on the transmission of radiation to the Earth's surface. A warmer atmosphere could hold more moisture, giving the possibility of increased cloudiness, and so less UV at the surface. The passage of weather systems may also be disturbed, creating local climate changes which will vary in their effect.

Tropospheric Ozone

The troposphere contains approximately 10% of the ozone in a vertical column, and while it may absorb UV radiation more effectively at low altitudes (because the radiation is largely scattered, giving it a longer pathlength through the tropospheric ozone), it is an unpleasant gas in proximity to the biosphere. Ozone is harmful to animal and human health, adversely affecting the respiratory system, and is also detrimental to some crops and plants. Thus air quality standards include limits for ozone, but these are frequently exceeded during the summer months in and around most major cities and industrial centres.

Ozone in the troposphere comes from the stratosphere, through downward incursions of stratospheric air, and through photochemical processes within the troposphere itself. Photochemical tropospheric ozone production/loss requires sunlight (UV) and is therefore linked to stratospheric ozone through the amount of UV radiation allowed to traverse the stratosphere and arrive in the lower atmosphere. The ozone photochemistry is detailed below:[16]

$$HO_2 + NO \rightarrow OH + NO_2$$
$$CH_3O_2 + NO + O_2 \rightarrow NO_2 + HO_2 + HCHO$$

followed by:

$$NO_2 + h\nu \rightarrow NO + O$$
$$O + O_2 \rightarrow O_3$$

and:

[14] D. T. Shanklin, D. Rind and P. Lonergan, *Nature*, 1998, **392**, 589.
[15] Stratospheric Ozone Review Group, *Stratospheric Ozone 1999*, The Stationery Office, London, 1999.
[16] Ultraviolet Measurements and Impacts Review Group, *The Potential Effects of Ozone Depletion in the United Kingdom*, The Stationery Office, London, 1996.

$$O_3 + h\nu \rightarrow O(1D) + O_2$$
$$O_3 + HO_2 \rightarrow OH + O_2 + O_2$$
$$O_3 + OH \rightarrow HO_2 + O_2$$
$$O_3 \rightarrow \text{dry deposition}$$
$$O_3 + NO \rightarrow NO_2 + O_2$$
$$O_3 + NO_2 \rightarrow NO_3 + O_2$$
$$NO_2 + NO_3 \rightarrow N_2O_5$$
$$N_2O_5 + H_2O \rightarrow 2HNO_3$$

It has been calculated[16] that for a global average change in ozone column depletion of -4.5%, the corresponding increase in UV will have its main impact on the production of singlet oxygen O(1D) and hence the hydroxyl free radical OH through

$$O_3 + h\nu \rightarrow O(1D) + O_2$$
$$O(1D) + H_2O \rightarrow OH + OH$$

In the free troposphere, that is well above the pollution layer associated with urban areas, the associated change in the tropospheric ozone column is calculated to be a depletion of between 1 and 2%. Close to the ground the additional UV from reduced ozone above can have the opposite effect, and on a local scale can produce undesirable increases in ozone. In terms of urban air quality, with ozone as a component of photochemical smog, the ozone production is limited by the availability of either hydrocarbons or NO_x. In the source regions of both hydrocarbons and NO_x (over cities) the production is hydrocarbon limited as there is enough NO_x to react with all the hydrocarbons and form ozone. In the countryside where NO_x becomes rapidly depleted through conversion to nitric acid, the hydrocarbons react with other peroxy radicals and ozone production is NO_x limited and ceases. Assuming no other changes, the increase in UV that accompanies loss of stratospheric ozone should increase the hydroxyl radicals OH and HO_2 and stimulate ozone production in the hydrocarbon limited regions of the lower troposphere, resulting in deteriorating air quality. There is little change expected near the ground in the rural (NO_x limited) regions. The overall trend in the contribution of tropospheric ozone to total column ozone is therefore difficult to determine. In some polluted areas, *i.e.* in and downwind of industrial regions, it may, overall, help to offset some of the stratospheric loss. In more pristine areas, and where technology is improving air quality, tropospheric ozone may decrease along with stratospheric ozone, adding to the increased transmission of UV to the surface.

5 UV Radiation at the Ground

The complexities, uncertainties and variable determinants of radiative transfer detailed above provide the background against which our knowledge and observations of UV radiation must be set. Natural variations in ozone and cloud both contribute to the widely fluctuating UV that can be considered within the climatological norm. Changes in ozone, cloud and pollution may occur singly or together, and act in concert or in opposition over different areas of the globe; determining whether they will have a detectable or significant effect on the UV climate, given its great natural range, is a challenge. One approach is to use radiative transfer models that allow calculations of the expected UV at the surface based on the physics of radiative transfer and a representation of the atmosphere in the model. Changing a single parameter (*e.g.* ozone) in the model atmosphere, which otherwise remains unchanged, allows estimates of the resulting changes in UV at the surface. However, the atmosphere and all its interactions are too complex to be completely represented in a model, and while models are a useful guide to expectations they can only make predictions for the relatively simple conditions on which they are based (usually cloud free, and, lacking other data, for a standard atmosphere). Ground-based observations of UV radiation measure the net result of all the factors affecting radiative transfer, but the observations themselves are not easy to make, do not necessarily identify the cause of any observed change in UV, and are not widely practiced. A global assessment of UV radiation at the surface must therefore rely on a combination of models and observations. This philosophy has been employed in recent satellite-derived assessments of global UV irradiance, using backscattered radiance data from satellites combined with a radiative transfer model to derive the surface UV irradiance.

6 Observations of UV Radiation

The ground-based UV observations available today are not the result of an organized global monitoring programme such as that for ozone; instead they are a haphazard mixture of national and individual measurement schemes using a range of different instruments and measurement protocols and unevenly distributed over the Earth's surface. The majority of monitoring sites are in North America and Western Europe,[17] while in the southern hemisphere Australia, New Zealand, South America and Antarctica have established, if sparser, UV monitoring networks. Elsewhere in the world, including the Tropics and Asia, monitoring sites are very scarce (Figure 2).

UV Instrumentation

The instruments used for UV monitoring fall into several categories: spectral instruments, broadband instruments and multi-channel instruments.[4,18,19] Spectroradiometers provide medium-high resolution spectral data that usually cover both the UV-B and UV-A spectral regions. The spectral detail allows more than just the simple UV fluxes to be extracted from the data: ozone and aerosol amounts can be calculated and the cause of any change in UV can be explored. Any change in UV due to ozone has a distinct spectral signature associated with the ozone absorption spectrum (Figure 3), and can be distinguished from the more wavelength-independent changes due to cloud, aerosol or surface albedo. All this information comes at a price and spectroradiometers have high capital

---

[17] E. C. Weatherhead and A. R. Webb, *Radiat. Prot. Dosim.*, 1997, **72**, 223.
[18] A. R. Webb, *UV-B Instrumentation and Applications*, Gordon and Breach, Reading, UK, 1998.
[19] World Meteorological Organisation, *Scientific Assessment of Ozone Depletion: 1998*, WMO Global Ozone Research and Monitoring Project report no. 44, WMO, Geneva, 1999.

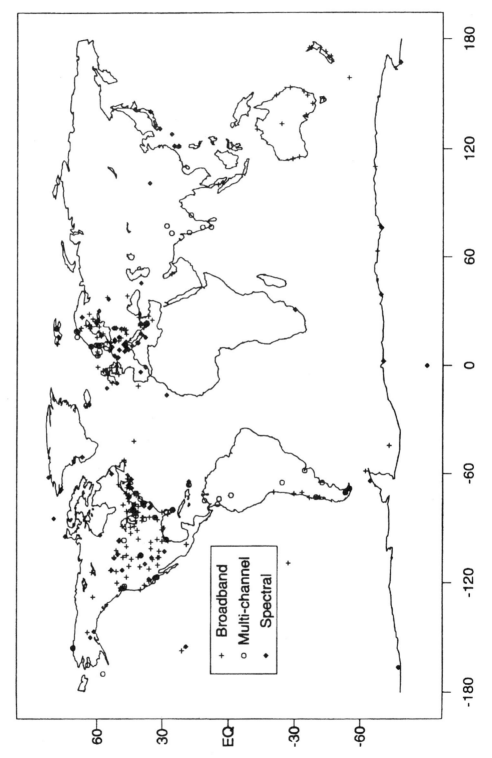

Figure 2 The global distribution of ground-based UV monitoring instruments (Updated from the 1997 data of Weatherhead and Webb[17])

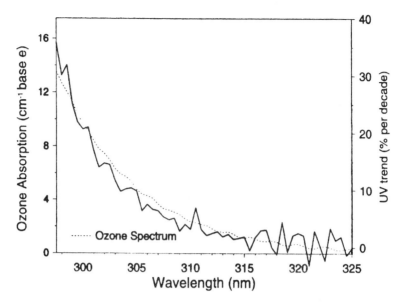

Figure 3 Spectral UV irradiance trends for May–August in Toronto, Canada. The similarity to the ozone absorption spectrum suggests that much of the UV change is due to changes in ozone (Taken from Wardle et al.[30])

and maintenance[20,21] costs which preclude their widespread use. Single or networked spectroradiometers were not deployed specifically to monitor UV until ozone depletion became apparent, and the earliest network was established in Antarctica by the U.S. National Science Foundation (NSF) in 1988.[22] Thus spectral data records are at best little more than a decade in length (a short time in climatological terms) and exist only at a few discrete sites. The most commonly deployed instruments are those in the broadband category. These radiometers measure the total incident radiation in a wide spectral band. By far the most common type of UV radiometer is one that measures the erythemally effective radiation, that is it has a response spectrum that mimics the erythemal (sunburn) action spectrum[23] across the UV-B and UV-A spectral regions. These radiometers evolved from concern about UV and skin cancer, sunburn being a risk factor for skin cancer, and were used in the earliest UV monitoring networks (which unfortunately have not been maintained through the period of ozone depletion). The radiometers are cheap and simple to operate and maintain in comparison to the spectroradiometers, though changes in the instrument's spectral response or sensitivity can be hard to identify and careful quality control is still necessary.[20] This ease of use means that the data from broadband radiometers provides the densest geographical coverage and longest time series of data available to the UV community. Multi-channel instruments are a comparatively new method of

---

[20] A. R. Webb, B. G. Gardiner, T. J. Martin, K. Leszczynski, J. Metzdorf and V. A. Mohnen, *Guidelines for Site Quality Control of UV Monitoring*, WMO/GAW 126, WMO, Geneva, 1998.
[21] B. Mayer and G. Seckmeyer, *J. Geophys. Res.*, 1999, **104**, 14321.
[22] C. R. Booth, J. C. Ehramjian, T. Mestechkina, L. W. Cabasug, J. S. Robertson and J. R. Tusson, *NSF Polar Programs UV Spectroradiometer Network 1995–1997 Operations Report*, Biospheric Instruments, San Diego, 1998.
[23] A. F. McKinlay and B. L. Diffey, *CIE J.*, 1987, **6**, 17.

measuring UV, combining some of the advantages of both spectral and broadband instruments. They measure the incident radiation in a series of narrow wavebands distributed across the UV-B and UV-A parts of the spectrum. Each channel is based on filter technology (like the broadband radiometer) and the channels can be sampled simultaneously, to give instantaneous if crude spectral detail—an advantage over the high-resolution spectroradiometer which often takes several minutes to produce a complete spectrum. The multi-band sampling allows cloud and aerosol effects to be separated from ozone effects by comparing changes in different parts of the spectrum. Multi-channel instruments have not been deployed to the same extent, or for as long, as broadband radiometers and spectroradiometers, but their use is increasing; for example, they are now used in networks in Scandinavia, South America and United States.

Data Archiving and Quality Control

In recent years, much effort has been expended in drawing together the many individual UV data records into archives that can provide a single source of compatible data. There are many problems with this task, not least the lack of a standard instrument or measurement protocol, even within each instrument category. The first task has been to explore the agreement, and causes for any lack of agreement, between instruments ostensibly measuring the same parameter, that is spectral or erythemally effective UV irradiance. This has been achieved through a series of instrument intercomparisons, the most comprehensive work coming from a series of European intercomparisons,[24] culminating in an intercomparison with the largest collection of spectroradiometers, from the widest global coverage, that took place in 1997.[25] The large discrepancies observed in the initial intercomparison in 1991 have been greatly reduced by identifying the causes of the discrepancies and working on both the instrumentation and measurement protocols to minimize each source of uncertainty. The limits to the agreement that can be expected, and to the uncertainty in any spectral UV data, is dependent on the stability and reproducibility of the standards of spectral irradiance against which the spectroradiometers are calibrated. The uncertainty in the reference sources available is generally 2–3% in the UV-B, added to which is the question of their long-term stability, the uncertainties generated by the transfer process (calibration), and the discrepancies that exist between standards from different National Standards Laboratories. These combined uncertainties in the underlying basis of all UV measurements mean that agreement of independently operated spectroradiometers to within 5% is an ambitious goal even for otherwise perfect instruments, yet today a significant number of spectroradiometers can, with careful operation, produce data consistently approaching such agreement with each other.

The largest comparison of broadband radiometers to date was in Helsinki in

---

[24] A. R. Webb (ed.), *Advances in Solar Ultraviolet Spectroradiometry*, Air Pollution Research Report 63, European Commission, Luxembourg, 1997.

[25] SUSPEN, *Standardisation of Ultraviolet Spectroradiometry in Preparation of a European Network*, ENV-CT95-0056, Final Report to the European Commission, Brussels, 1998.

1995,[26] when it was found that careful characterization of each instrument, and a common calibration, greatly improved the agreement between instruments. The calibration of broadband instruments is made against the spectrally weighted simultaneous measurements of a spectroradiometer, using the sun as a source. The uncertainties in the spectroradiometer measurements then form the basis of uncertainties in the broadband data. Added to this are further uncertainties in the calibration process, imperfections in both instruments, and the fact that owing to the changing solar spectrum with SZA combined with the spectral response of both radiometers, a calibration is not truly valid for any atmospheric conditions other than those under which it was performed. As with the spectroradiometric measurements, some of the sources of uncertainty can be minimized if the instrument is well characterized and the conditions under which the measurements were made are known. Nonetheless, it is fortuitous if independent broadband instruments agree to within 10%.

The absolute agreement between instruments at an intercomparison, a discrete event in time, is one measure of quality assurance. However, this performance must also be maintained at the home site over long periods of time. When searching for small changes in a measured parameter that has the large natural variability of UV irradiance, stability is the most important factor: to directly detect trends in UV-B expected with the observed (small) trends in ozone at mid-latitudes, a stability of 1–2% per decade is required. Continuous quality control on-site is therefore necessary to keep the relative as well as absolute uncertainties in the data as small as possible and enable any changes or trends in the data to be identified.[20,27] Further quality control may be performed as data enter a central data archive and it is important that the estimated uncertainty in a data set is provided with the data so as not to mislead any data users, especially when there is no definitive standard for UV instrumentation or data.

There are currently two centralized data archives for UV data, in addition to the individual site or network databases that exist. One is the World Ultraviolet Data Centre (WUDC) of the World Meteorological Organisation, held in Toronto alongside the World Ozone Data Centre.[28] This will accept UV data from any source, spectral, broadband or multi-channel, and access is widely available. The other database is the European UV Database, established with European Community funding and physically held at the Finnish Meteorological Institute in Helsinki.[29] This database concentrates on spectral data gathered within the European Community, although some broadband and ancillary data are also available. Access to the data is available to those outside the European project through a Cooperation Agreement.

---

[26] K. Leszczynski, K. Jokels, L. Ylianttila, R. Visuri and M. Blumthaler, *Photochem. Photobiol.*, 1998, **67**, 212.

[27] G. Seckmeyer, A. Bais, G. Bernhard, M. Blumthaler, C. R. Booth, P. Disterhof, P. Eriksen, R. L. McKenzie, M. Miyauchi and C. Roy, *Instruments to Measure Solar Ultraviolet Radiation Part 1: Spectral Instruments*, WMO/GAW 125, WMO, Geneva, 2000.

[28] http://www.ec.gc.ca/woudc

[29] http://www.ozone.fmi.fi/SUVDAMA/index.html

Analysis of Ground-based Observations

Evidence of any changes in UV radiation measured at the ground comes from a small number of sites and these local observations cannot be applied to wider regions. Nonetheless, the available data sets tell a consistent story of anti-correlation with ozone, shown through a variety of analysis techniques.

The most obvious changes in UV radiation occur where ozone depletion is greatest: in Antarctica, and more recently in the Arctic, springtime increases in UV-B have been easily detectable.[22] The total dose at any site for a given period depends upon latitude (SZA and daylength), cloud and character of the surface as well as ozone amount: maximum doses are found when low ozone coincides with clear skies at a time of relatively long daylength, small SZA and highly reflecting snow surface. Thus the prolongation of Antarctic ozone depletion into the late spring/summer months can greatly increase UV-B doses. At present the maximum daily average springtime doses at Palmer Station (64.8°S) exceed those of San Diego (32.8°N) in mid-summer. The large year-to-year variability in timing, position and depth of the ozone hole, and cloud at a given station, make it harder to identify trends but there is evidence of continuing increases in UV-B. The spectral data from the first established monitoring site at Palmer Station shows the most compelling trend for daily averages of DNA-weighted UV (the solar spectrum weighted with the DNA action spectrum) during the month of November. The site has been in operation since 1988, and since that time shows a trend of +15% per year with a correlation coefficient of 0.86. The change in visible radiation (not affected by ozone) for the same period was +2.7%.[22]

At mid-latitudes, where long-term ozone depletion is only a few percent per decade and cloud and aerosol variability acts on many time scales, changes in UV-B are harder to identify directly, especially as data records are generally less than decadal in length and the long-term stability of the instrumentation has yet to be clearly demonstrated. Where spectral data are available the shorter (more ozone dependent) wavelengths can be studied as they should show the greatest response to ozone changes, though they also provide the greatest measurement challenge. One of the longest spectral data records, beginning in 1989, comes from Toronto, Canada. Daily average UV fluxes for individual wavelengths have been calculated for the summer months (May to August) when UV irradiances are greatest, giving a trend of +15% per decade at 300 nm reducing to zero trend at 324 nm, while the corresponding decrease in ozone was −4.3% per decade.[30] The spectral trend follows the ozone absorption spectrum closely (Figure 3), clearly identifying the cause of the increased UV.

The absolute UV irradiance at a single wavelength well absorbed by ozone (305 nm) has been reported from four stations in Europe (Thessaloniki, Brussels, Garmisch-Partenkirchen and Reykjavik) with data records between 3 and 6 years in length.[31] Monthly mean values of the irradiances at a SZA of 63° showed UV changes of between +9% and −3% per year for corresponding ozone changes

---

[30] D. I. Wardle, J. B. Kerr, C. T. McElroy and D. R. Francis, *Ozone Science, a Canadian Perspective on the Changing Ozone Layer*, Environment Canada, Toronto, 1997.

[31] C. S. Zerefos, D. S. Balis, A. F. Bais, D. Gillotay, P. Simon, B. Mayer and G. Seckmeyer, *Geophys. Res. Lett.*, 1997, **24**, 1363.

of -1.5% and $+0.5\%$ per year at three of the stations, although the effect of increasing the length of the data record was shown to have a significant effect on the calculated trends for these short data records and the figures should not be extrapolated to infer long-term trends. At the fourth station (Garmisch-Partenkirchen) the effect of cloud on a short record of absolute UV irradiances was illustrated. An exceptionally cloud-free October distorted any attempt at trend analysis in a short-term data record.

Cloud and other meteorological variables can be removed from the trend analysis, or their effect reduced, by taking a ratio of UV-B and UV-A wavelengths. So long as there is no strongly wavelength dependent factor other than ozone acting over the period of measurement (*e.g.* the wavelength dependency of the instrument remains stable), the ratios should identify any relative trends in UV-B radiation, but they do not provide information about changes in absolute UV-B irradiance.[32,33] Such an analysis has been performed on 6 years (1993–1998) of spectral data from Reading, UK.[32] The monthly mean ratios of erythemal UV to UVA were calculated for SZAs between 75° and 80° (the solar spectrum and therefore the ratios are SZA dependent) and assessed with respect to the corresponding column ozone. There was a clear anti-correlation between the seasonal cycles of ozone and UV ratios, and for -2% change in ozone over the period there was an increase in the erythemal UV ratio of $+3.5\%$, although the variability of the data makes this small change statistically insignificant.

The longest time series of UV measurements come from broadband instruments, which do not allow for the sort of spectral analysis detailed above (and therefore include cloud effects and any changes in pollution), and are not as sensitive to ozone change as the shortest UV wavelengths. Data from Russia (1968–1983), Poland (1976–1994), USA (1974–1985, revised data), USA (1974–1987) and UK (1988–1997) all show increases in UV radiation over the period of measurement, but owing to the large variability in the data and uncertainties in the long-term stability of the instruments, some of which were changed during the monitoring periods, the trends are generally not statistically significant.[7,19] Broadband measurements made at altitude in the Swiss Alps (avoiding the influences of pollution) since 1981 have shown small increases in erythemal UV, consistent with ozone changes at the same site. There have also been reports from an increasing number of sites of short-term enhancements in UV irradiance coinciding with episodes of extreme low ozone over a site.[19] While such events do not constitute a trend, their increasing frequency may indicate on-going atmospheric changes that can be identified in the UV records.

As UV monitoring sites, and the length of data records, increase, the evidence for changing UV irradiances at the surface continues to accumulate, albeit from limited locations. To obtain a view of what is happening on a more global scale a different approach is needed.

---

[32] L. M. Bartlett and A. R. Webb, *J. Geophys. Res.*, 2000, **105**, 4889.
[33] A. F. Bais, M. Blumthaler, A. R. Webb, J. Groebner, P. J. Kirsch, B. G. Gardiner, C. S. Zerefos, T. Svenoe and T. J. Martin, *J. Geophys. Res.*, 1997, **102**, 8731.

A. R. Webb

Satellite-derived UV Irradiances

The Total Ozone Mapping Spectrometer (TOMS) satellite instruments have operated from 1978–1993, 1991–1994 and 1996–present, providing a 20-year data record (although intercalibration issues with the latest instrument must be resolved before its data can be added to the 1978–1994 time series).[19] Global daily estimates of UV irradiances are derived from the TOMS data using the measured ozone amounts, cloud reflectivities, aerosol amounts and scene reflectivities combined with a radiative transfer model. The 1978–1994 TOMS data show statistically significant changes in ozone at all latitudes larger than 35°, and no trend in zonally averaged cloud reflectivity. The zonally averaged trend in erythemal irradiance derived from the measurements was $2.9 \pm 3\%$ per decade for both latitude bands between 35° and 45°, while at 60°N the estimated trend was $3.7 \pm 3\%$ and at 60°S it was $9 \pm 6\%$ per decade.[34] The satellite derivations agree well with ground-based measurements from a single station (Toronto), though there are seasonal differences in the level of agreement, but further validation is required against other ground-based measurements.

7 Longer-term Assessments of UV Irradiances

Putting the current observed changes in UV irradiances into context by considering past levels of UV flux, and future expectations, relies on the use of UV radiative transfer models and estimates of the past or future state of the atmosphere. There are many types of radiative transfer model, of varying complexity, available for model studies. When using the same input data (model atmosphere), many of the models agree reasonably well. Models also show good agreement with UV measurements in clear sky conditions when there is enough ancillary data available to make a good approximation to the real atmosphere.[19] The largest uncertainties in modelled UV irradiances come from uncertainties (mismatches with reality) of the input data,[35] and these uncertainties increase greatly in non-clear conditions; the complexity of real cloud is poorly represented in models. However, models can be used to estimate future changes in UV for a specified change in one (or more) input parameters, assuming that nothing else changes, or to calculate the past UV climate from what is known or assumed about the atmosphere of the past. While these calculations may not provide good estimates of the absolute UV levels to be expected (they usually neglect cloud for simplicity), they do give a prediction of the relative change incurred by, for example, a 10% ozone depletion.

Recent calculations of the changes in UV radiation since pre-industrial times owing to stratospheric and tropospheric ozone changes[36] were based on a current baseline climatology using model inputs of three-dimensional monthly mean climatology of ozone and temperature, cloud and surface albedo observations, and a fixed aerosol amount.[37] The current climatology shows that the most

---

[34] J. R. Herman, P. K. Bhartia, J. Kiemke, Z. Ahmad and D. Larko, *Geophys. Res. Lett.*, 1996, **23**, 2117.
[35] P. Weihs and A. R. Webb, *J. Geophys. Res.*, 1997, **102**, 1541.
[36] A. A. Sabziparvar, K. P. Shine and P. M. Forster, *J. Geophys. Res.*, 1998, **103**, 26107.
[37] A. A. Sabziparvar, P. M. Forster and K. P. Shine, *Photochem. Photobiol.*, 1999, **69**, 193.

important factor in determining geographical distribution of daily UV is the sun's position, rather than column ozone. In regions with high surface elevation the effect of altitude was more important in determining UV than differences in ozone, while clouds reduce UV by up to 45% in regions experiencing the frequent passage of low-pressure systems. Against this baseline climatology the estimated pre-industrial UV climate shows that at low latitudes (where there has been no significant change in stratospheric ozone) the daily integrated erythemally effective UV may have decreased by up to 9% owing to increases in tropospheric ozone (from increased biomass burning and industrialization), while at high latitudes UV has increased significantly during the periods of currently observed ozone depletion. Over much of the rest of the northern hemisphere, summertime values of erythemal UV are seen to decrease from the pre-industrial level, with increased tropospheric ozone and only small stratospheric ozone depletion. Crude estimates were also made of the possible effects of aerosol concentrations on the long-term changes in UV. The possible global reduction in UV due to aerosols was calculated as 2%, although locally the effects could be much greater, for example soot aerosol could reduce UV by more than 6%. While there is significant uncertainty in the long-term changes estimated in this way, they do help to put the trends due to recent ozone depletion into perspective, and show that changes in stratospheric ozone are not the only cause of UV trends.

Future predictions of UV irradiances are based primarily on predicted changes in ozone, which in turn are model results dependent on estimates of the future atmospheric loading of ozone-depleting substances and our current understanding of ozone-depleting processes. There are ozone models of varying complexity, from two-dimensional (latitude–height) models to three-dimensional (latitude–longitude–height) coupled chemistry–climate models. Different two-dimensional models show similar patterns of ozone recovery, though the details differ, and infer that ozone will not return to pre-1980 levels before 2050, and by that time the vertical distribution of ozone is likely to differ from that today because of other changes in atmospheric constituents. The more complex chemistry–climate models take into account the fact that ozone will be recovering to a climate that is not the same as that today (altered by greenhouse warming), and the altered temperature structures will also influence the ozone depletion processes. These models, run for the polar regions, differ widely in some aspects, but generally show that the recovery of the Arctic ozone is likely to be delayed by the increase of polar stratospheric clouds due to greenhouse gases. In the Antarctic, similar processes prevent a steady increase in ozone after the minimum (predicted around 2010–2015) in some of the models.[15] The return of UV to the pre-ozone depletion norm should coincide with ozone recovery, if other climate changes (*e.g.* changed cloud) are not significant.

8 UV Forecasting

Daily weather forecasts in several countries include a forecast for UV, either routinely or during the summer months (as in the UK) when UV is of a magnitude to induce sunburn. The UV index has been internationally agreed as the means of disseminating UV exposure information to the public, and ranges from a value of

1 (essentially no danger of sunburn) to 12 or more for tropical conditions. The index is based on measured or calculated UV irradiance on a horizontal surface, and so must be applied with care to surfaces of other orientations. Several methods are used for UV forecasts, but they all essentially forecast near-future ozone from current observations and then calculate the UV, including a cloud amount prediction in some cases to assess the cloud attenuation effect. Accurately predicting the cloud fields, and then their influence on the local UV irradiance, is the most challenging aspect of short-term UV forecasting.

9 Conclusion

Changes in UV irradiances at the Earth's surface associated with stratospheric ozone depletion have been directly observed (at a limited number of locations), derived from satellite observations on a global scale, and calculated from climatological data. There are uncertainties in all the methods of UV trend detection or estimation, and outside polar regions the changes are small (as are the changes in stratospheric ozone). Ozone is not the only determinant of UV at the surface, and on a longer time scale atmospheric changes other than stratospheric ozone may have been, or will be, more influential than current ozone depletion at low and mid-latitudes. The future of the stratospheric ozone layer is uncertain in the detail of its recovery, but it is expected to recover as ozone depleting substances in the atmosphere are reduced. With this recovery should come a reduction in UV at the surface, moderated however by any changes that there might be in aerosols, tropospheric ozone and cloud amount. All these changes in UV must be viewed against the extensive range and variability of UV irradiances which naturally occur at any location, and the significance of any changes in terms of UV effects will depend upon the mechanisms by which each effect acts and its susceptibility to change.

10 Acknowledgements

The author would like to thank Betsy Weatherhead for providing the updated version of Figure 2, and Berit Kjeldstad for useful comments on the manuscript.

Marine Photochemistry and UV Radiation

ROBERT F. WHITEHEAD AND STEPHEN DE MORA

1 Introduction

Solar radiation, directly or indirectly, is the driving force for most physical, biological and chemical processes in the ocean. Included amongst those directly affected by solar radiation are marine photochemical reactions. Photochemistry is not a new field of study and there are many excellent texts and reviews of its foundations.[1,2] The incorporation of photochemical studies into the field of environmental and marine chemistry is, however, a relatively new endeavour. Over the past 20 years, it has become clear that photochemical reactions can play a significant role in the overall chemical milieu of the sun-lit surface ocean. There have to date been several reviews and symposia of natural water and seawater photochemistry.[3-9] The continuing efforts to elucidate specific photoprocesses in natural waters have been documented in a number of more specialised reviews concerning photochemical reaction rates and quantum yields,[10-12] production of

---

[1] J. Calvert and J. Pitts, *Photochemistry*, Wiley, New York, 1966.
[2] N. J. Turro, *Modern Molecular Photochemistry*, Benjamin/Cummins, Menlo Park, 1978.
[3] O. C. Zafiriou, *Mar. Chem.*, 1977, **5**, 497.
[4] O. C. Zafiriou, in *Chemical Oceanography*, ed. J. Riley and R. Chester, Academic Press, London, 1983, p. 339.
[5] R. Zika, in *Marine Organic Chemistry*, ed. E. Duursma and R. Dawson, Elsevier, Amsterdam, 1981, p. 299.
[6] O. C. Zafiriou, J. J. Dubien, R. G. Zepp and R.G. Zika, *Environ. Sci. Technol.*, 1984, **18**, 358A.
[7] R. Zika and W. Cooper, *Photochemistry in Environmental Aquatic Systems*, American Chemical Society, Washington, 1987, p. 281.
[8] N. V. Blough and R. G. Zepp, *Effects of Solar Ultraviolet Radiation on Biogeochemical Dynamics in Aquatic Environments*, Woods Hole Oceanographic Institute Technical Report WHOI-90-9, Woods Hole, 1990.
[9] G. R. Helz, R. G. Zepp and D. G. Crosby, *Aquatic and Surface Photochemistry*, Lewis, Boca Raton, 1994, p. 491.
[10] R. G. Zepp and D. M. Cline, *Environ. Sci. Technol.*, 1977, **11**, 359.
[11] W. Miller, *Ecol. Stud.*, 1998, **133**, 125.
[12] T. Mill, *Chemosphere*, 1999, **38**, 1379.

radicals and reactive oxygen species,[13-16] formation of biologically labile compounds[17,18] and influences on biogeochemical cycles.[19,20] A vast majority of these studies and those cited therein indicate that it is light in the near ultraviolet (UV) and the short visible wavelengths of the solar spectrum that is primarily responsible for most marine photochemical reactions. This observation, coupled with the recorded declines in stratospheric ozone,[21] has spurred interest in the potential effects of increased ultraviolet-B radiation[22] (UV-BR, 280–320 nm) on marine photobiological and photochemical processes.[23]

2 Basics of Marine Photochemistry

Light and Absorption

Marine photochemistry is itself merely a specialised branch of photochemistry, with the general condition that only light wavelengths found in the solar spectrum are of interest. As such, a brief outline of the basics of photochemistry is given for the benefit of readers with a limited background in the subject. While these basics apply to all aquatic photochemical processes, it should be noted that marine photochemistry processes are dominated by those involving solar UV radiation. Some of the difficulties particular to defining and interpreting marine photochemical processes are also discussed.

Solar radiation, and light in general, may be regarded as having both wave-like and particle-like properties. From a photochemical point of view, it is often easiest to visualise processes when light is considered as a particle. In this context, light is emitted, transmitted and absorbed in discrete units quantified as photons or quanta. The energy (E) of a photon in Joules is described by:

$$E = \frac{hc}{\lambda} \tag{1}$$

where h is the Planck constant (6.63×10^{-34} J s), c is the speed of light in a vacuum

---

[13] R. G. Zepp, in *Humic Substances and Their Role in the Environment*, ed. F. H. Frimmel and R. F. Christman, Wiley, New York, 1988, p. 193.

[14] W. Cooper, R. Zika, R. Petasne and A. Fischer, in *Aquatic Humic Substances: Influence on Fate and Treatment of Pollutants*, ed. I. Suffet and P. McCarthy, American Chemical Society, Washington, 1989, p. 333.

[15] J. Hoigné, B. C. Faust, W. R. Haag, F. E. Scully and R. G. Zepp, in *Aquatic Humic Substances: Influence on Fate and Treatment of Pollutants*, ed. I. Suffet and P. MacCarthy, American Chemical Society, Washington, 1989, p. 363.

[16] N. V. Blough and R. G. Zepp, in *Active Oxygen in Chemistry*, ed. C. S. Foote, J. S. Valentine, A. Greenberg and J. F. Liebman, Chapman and Hall, New York, 1995, p. 280.

[17] M. A. Moran and R. G. Zepp, *Limnol. Oceanogr.*, 1997, **42**, 1307.

[18] D. J. Kieber, in *Effects of UV Radiation on Marine Ecosystems*, ed. S. J. de Mora, S. Demers and M. Vernet, Cambridge University Press, Cambridge, 2000, p. 130.

[19] R. G. Zepp, T. V. Callaghan and D. J. Erickson, *Ambio*, 1995, **24**, 181.

[20] K. Mopper and D. J. Kieber, in *Effects of UV Radiation on Marine Ecosystems*, ed. S. J. de Mora, S. Demers and M. Vernet, Cambridge University Press, Cambridge, 2000, p. 101.

[21] P. J. Crutzen, *Nature*, 1992, **356**, 140.

[22] S. Madronich, R. L. McKenzie, M. M. Caldwell and L. O. Björn, *Ambio*, 1995, **24**, 143.

[23] R. F. Whitehead, S. J. de Mora and S. Demers, in *Effects of UV Radiation on Marine Ecosystems*, ed. S. J. de Mora, S. Demers and M. Vernet, Cambridge University Press, Cambridge, 2000, p. 1.

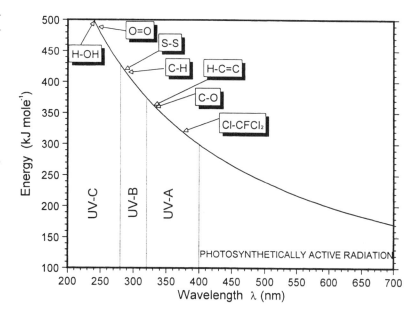

Figure 1 A comparison of the inverse relationship of light energy with wavelength. The position of some selected bond energies on the curve are indicated by the arrows. In principle, absorbed light can cleave a chemical bond with equal or lesser energy. The energies of many bonds in marine organic material are equivalent to light energies in the UV region

(3.0×10^{17} nm s^{-1}) and λ is the wavelength (nm). The term Einstein is often used to describe quanta on a molar basis such that 1 Einstein (or mol photon) is equivalent to 6.023×10^{23} photons. From this relationship, it is clear that the energy of a photon is inversely proportional to its wavelength. A comparison of light energies with energies of bonds commonly encountered in natural systems reveals that overlaps generally occur in the UV and short visible wavelength ranges (Figure 1). Although other factors must be considered, in principle, absorption of light with sufficient energy is able to cleave bonds with equal or lesser energy.

Whether or not a photochemical reaction takes place is governed by two fundamental laws of photochemistry:

1. Only light which is absorbed can effect change in a chemical system (Grotthus–Draper)
2. Only one molecule is activated for each photon absorbed in a system (Stark–Einstein).

The first law is intuitive in that light must be absorbed in order for its energy to be transferred to a molecule. In environmental photochemistry, absorption of photons is determined by the overlap of the solar radiation spectrum and the absorption spectrum of the compound of interest. The direct photolysis of chlorofluorocarbons (CFCs) demonstrates the importance of considering this overlap. Although light with wavelengths as long as ~ 360 nm has sufficient energy to cleave a Cl atom from a CFC molecule (Figure 1), absorbance by CFCs decreases rapidly beyond 200 nm and is virtually zero by 290 nm. This weak absorbance of longer wavelengths combined with the strong attenuation of short wavelength light in the upper atmosphere severely limits direct photolysis of

CFCs in the troposphere. In the upper atmosphere, however, there is sufficient short wavelength light to yield effective direct photolysis of CFCs, thereby facilitating subsequent reactions with stratospheric ozone.

The second law requires that an absorbed photon be removed from the system and thus yields a proportional relationship between primary photochemical reactions and the number of photons absorbed. It should be noted that, owing to instrument limitations in spectral resolution, irradiance is often measured in terms of energy (*i.e.* $J m^{-2} s^{-1} = W m^{-2}$) over broad wavelength bands. Photolysis rates based on these data and given in change per unit energy absorbed per unit time are therefore not strictly quantitative.

Considering the basic laws of photochemistry, it is obvious that absorption of incident solar radiation is the first and fundamental event in environmental photochemistry. The absorbance of radiation is quantified using the familiar Lambert–Beer law, which relates light intensity (I) emerging from a solution after passing through a known pathlength to the incident light intensity (I_0):

$$I(\lambda) = I_0(\lambda) \times 10^{-[\alpha(\lambda) + \varepsilon(\lambda)C]l} \qquad (2)$$

or absorbance $A_\lambda \equiv \log[I_0(\lambda)/I(\lambda)] = [\alpha(\lambda) + \varepsilon(\lambda)C]l \qquad (3)$

C is concentration of the compound of interest in mol L^{-1}, l is pathlength in cm, $\alpha(\lambda)$ is the decadic absorption or attenuation coefficient (cm^{-1}) of the solvent (*i.e.* seawater for marine situations), and $\varepsilon(\lambda)$ is the decadic molar extinction coefficient ($L\ mol^{-1}\ cm^{-1}$) of the compound at wavelength λ. Although water is a strong absorber at the red and infrared end of the spectrum, pure water contributes little to absorbance in the near UV.[24] In addition, the inorganic salts in seawater are not major contributors to light absorption of wavelengths in the solar spectrum.[6] However, in unfiltered seawater samples, both inorganic and organic suspended particles may contribute significantly to both scattering and absorbance.[24] With the appropriate reference solution (commonly de-ionised, carbon-free water), the absorption spectrum of the dissolved constituents of filtered seawater can be readily determined using a spectrophotometer. Although influenced by numerous factors, including but not limited to atmospheric ozone concentrations, the practical lower limit for solar radiation reaching the sea surface is approximately 290 nm. Therefore, in order to *initiate* a photochemical change, a seawater constituent must have some ability to absorb radiation in the UV-B (280–320 nm) or longer wavelength ranges.

In marine environments, dissolved organic material (DOM) is the principal dissolved component responsible for absorption of solar radiation in the UV and short visible wavelengths.[24,25] DOM is present in surface seawater at concentrations of about 60–300 μmol L^{-1} C (0.7–3.6 mg C L^{-1})[26] and represents one of the largest pools of reduced carbon at the Earth's surface.[27] Only a small percentage of marine DOM, 10–50%, is readily identifiable as common biomolecules such as

[24] J. T. O. Kirk, *Light and Photosynthesis in Aquatic Ecosystems*, Cambridge University Press, Cambridge, 1994.
[25] A. Bricaud, A. Morel and L. Prieur, *Limnol. Oceanogr.*, 1981, **27**, 43.
[26] T. F. Thingstad, A. Hagstrom and F. Rassoulzadegan, *Limnol. Oceanogr.*, 1997, **42**, 398.
[27] J. I. Hedges, *Mar. Chem.*, 1992, **39**, 67.

Figure 2 A comparison of the absorption spectra of CDOM from different areas of the St. Lawrence Estuary, Quebec, Canada with modelled downwelling irradiance of a mid-latitude summer and winter day at noon. Although the shapes of CDOM absorption spectrum from any marine water sample are generally uniform, the magnitude is highly variable and is a quasi-function of DOC concentration and salinity in many waters

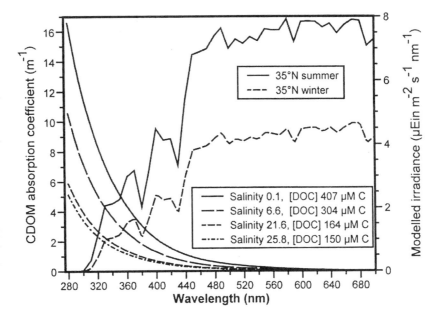

amino acids, polysaccharides, *etc.*[28] The dominant uncharacterised fraction is composed of complex heteropolycondensates derived from *in situ* and terrestrial sources.[29] DOM concentrations are usually greatest in coastal waters, whereas open ocean waters have lower concentrations and show very little terrestrial influence.[30] The uncharacterised fraction has been variously termed marine humic material, yellow substance, gelbstoffe or gilvin. This fraction apparently contains the majority of 'unknown photoreactive chromophores'[6] and is currently referred to as chromophoric dissolved organic material (CDOM). Owing to its complexity and variability, an accurate universal molar extinction coefficient cannot be defined for CDOM. Thus, to provide consistency to absorbance measurements taken using various pathlengths, absorbance (A, unitless) measured with a spectrophotometer is frequently converted to an absorptivity (cm^{-1}) by using the relationship:[24]

$$a_{\text{CDOM }\lambda} = \frac{2.303 A_\lambda}{l} \tag{4}$$

where l is the pathlength in centimetres. Although the magnitude of a_{CDOM} varies from one site to another, the shapes of CDOM absorptivity (a_{CDOM}) spectra from most marine samples (and most natural water samples) are remarkably similar, exhibiting a generally featureless increase in absorbance from visible to UV wavelengths (Figure 2). As with DOM concentrations, highest a_{CDOM} is commonly

[28] E. M. Thurman, *Organic Geochemistry of Natural Waters*, Junk, Boston, 1985.
[29] J. Hedges, in *Humic Substances and Their Role in the Environment*, ed. F. Frimmel and R. Christman, Wiley, New York, 1988, p. 45.
[30] S. Opsahl and R. Benner, *Nature*, 1997, **386**, 480.

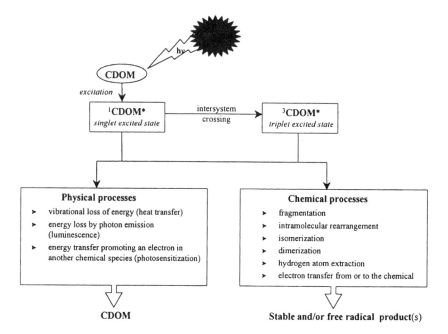

Figure 3 A simplified diagram showing the various photophysical and photochemical processes that can occur following the absorption of sunlight by marine CDOM. Physical processes (left side) return excited CDOM to ground state without causing a chemical change in the system. Chemical processes (right side) are responsible for the wide array of radical and non-radical species observed in irradiated seawater

found in coastal waters with strong terrestrial influence and lowest a_{CDOM} is found in clear open-ocean waters.[25,31–33]

Primary and Secondary Reactions

When light of sufficient energy is absorbed by CDOM (or any absorbing compound), the energy transferred to the compound can elevate an electron from a lower electronic state to a higher electronic level. The most commonly produced singlet excited species (^{1}CDOM*) will subsequently lose the acquired energy through various photophysical or photochemical pathways (Figure 3). A large portion of excited species will lose the absorbed energy through vibrational or rotational relaxation, internal conversion, re-emission of a photon (fluorescence) or collision deactivation. These processes yield a return to the ground state for the excited species without effecting a chemical change in the system. However, some collision deactivations may transfer sufficient energy to an acceptor compound to produce a secondary excited species. Further reactions involving these secondary excited species are termed photosensitised or indirect photoreactions. Some portion of singlet excited species may undergo intersystem crossing to triplet excited species (^{3}CDOM*). Although triplet excited species can themselves relax through the same photophysical pathways as singlet excited species, they are at a lower energy level than singlet species and consequently have longer lifetimes. A large percentage of photochemical reactions in seawater is thought to involve these longer-lived triplet excited species.[11] A more thorough discussion of

[31] J. R. Nelson and S. Guarda, *J. Geophys. Res.*, 1995, **100**, 8715.
[32] N. V. Blough, O. C. Zafiriou and J. Bonilla, *J. Geophys. Res.*, 1993, **98**, 2271.
[33] S. A. Green and N. V. Blough, *Limnol. Oceanogr.*, 1994, **39**, 1903.

photophysical processes can be found in text such as Calvert and Pitts[1] or Turro.[2]

In addition to the primary photophysical processes, both singlet and triplet excited species may undergo a variety of primary photochemical reactions (Figure 3). These pathways result in the formation in a variety of stable and free radical species. The products of the primary photochemical reactions may then further react through secondary photochemical, chemical or biological processes. Owing to the complexity of CDOM and the marine chemical milieu, quantifying or even identifying all possible photochemical reaction products has thus far proven impossible. Those photoproducts that have been studied are the focus of the next section of this review.

Quantum Yields and Reaction Kinetics

Primary photophysical and photochemical processes can be quantitatively assessed in terms of the relative efficiency of each particular pathway (*i.e.* fluorescence, isomerisation, *etc.*). This efficiency is termed the quantum yield (Φ) and the sum of quantum yields of all *primary processes* is one. Unfortunately, there are no reliable guidelines for estimating quantum yields and they must be determined experimentally. In environmental photochemistry, determination of quantum yields for each individual pathway is currently not feasible. Instead, a reaction quantum yield can be defined as:

$$\Phi_{r\lambda} = \frac{\text{Moles of product formed or reactant lost}}{\text{Moles of photons absorbed by the reactant}} \quad (5)$$

The reaction quantum encompasses all pathways that yield the product or transform the reactant of interest. Because CDOM is not well characterised and owing to the probability of secondary reactions in most natural waters, the term apparent quantum yield $\Phi_{a\lambda}$ is used when the reaction quantum yield is normalised to the total absorbance of dissolved material, rather than the absorbance of a specific known reactant. Although $\Phi_{a\lambda}$ may exceed one in cases where the primary photochemical process initiates a chain reaction, in natural waters the values for $\Phi_{a\lambda}$ are generally much lower than unity owing to competition amongst the various decay pathways after a photon has been absorbed. Several reviews on the calculation of quantum yields have been published.[10,11,34,35]

Quantum yields are generally assumed to be wavelength independent when describing a specific process induced by the absorption of a photon within a specific electron transition in a specific compound. The poorly characterised nature of the electronic transitions in CDOM prevents the use of such a rigorous definition and requires that the wavelength dependence of $\Phi_{a\lambda}$ be considered. In fact, the relatively few quantum yields reported for natural waters do display wavelength dependence (Figure 4), with highest values observed in the UV region. Additionally, some critical wavelength occurs above which photons do

[34] J. Lemaire, I. Campbell, H. Hulpke, J. Guth, W. Merz, J. Philip and C. von Waldow, *Chemosphere*, 1982, **11**, 119.

[35] W. Draper, in *Photochemistry of Environmental Aquatic Systems*, ed. R. Zika and W. Cooper, American Chemical Society, Washington, 1987, p. 268.

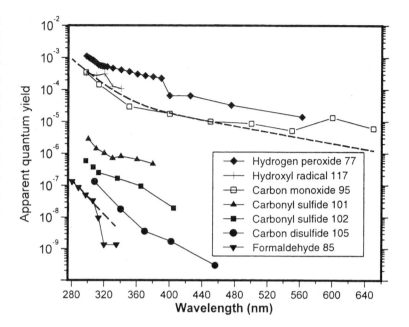

Figure 4 Reported apparent quantum yield spectra for various marine and freshwater photoproducts. References are given for each compound. Negative slopes of the spectra are evident even on a logarithmic scale, which illustrates the heavy weighting of UV wavelengths in most marine photochemical processes. The dashed lines shown for carbon monoxide and formaldehyde are the least squares regression lines that are used in the model described in the text

not have enough energy to generate an excited species and thus quantum yields will fall to zero. By definition, quantum yields are a function of the number of photons absorbed and not light intensity. However, apparent quantum yields include both secondary reactions and total absorbance, some of which may not be specific to the reaction, and therefore may be related to light intensity in complex ways.

Plots of quantum yields versus wavelength provide a graphical representation of a photochemical response function. Action spectra for photochemical reactions can be produced from plots of the product of $\Phi_{a\lambda}$ times a_λ *versus* wavelength. Multiplying an action spectrum by an incident irradiance yields a plot of maximum photochemical reaction rates *versus* wavelength that facilitates comparisons among the expected responses for different water samples or light spectra. A rigorous treatment of photochemical reaction rates requires careful consideration of light attenuation and absorbance in the system. In the simplest terms, a *unimolecular primary* photochemical reaction rate is related to the quantum yield by:

$$-(dC/dt) = \Phi_\lambda I_{a\lambda} \qquad (6)$$

where C is concentration in mol L^{-1} and $I_{a\lambda}$ is the specific light absorbance rate in mol quanta L^{-1} s^{-1}. The specific light absorbance rate is related to the incident irradiance ($I_{0\lambda}$ in mol quanta cm^{-2}) by:

$$I_{a\lambda} = I_{0\lambda}(1 - 10^{-[\alpha(\lambda) + \varepsilon(\lambda)C]l})\left(\frac{\text{area}}{\text{volume}}\right)\frac{\varepsilon(\lambda)C}{\alpha(\lambda) + \varepsilon(\lambda)C} \qquad (7)$$

where the terms in parentheses are as defined in equation (2), area and volume are the surface area of the light beam and volume of the irradiated sample, respectively, and the last term represents the fraction of light absorbed by the photoreactive compound.

When a system is optically thin (i.e. $[\alpha(\lambda) + \varepsilon(\lambda)C]l < 0.02$), only a very small percentage of incident radiation is absorbed and equation (7) simplifies to:

$$-(dC/dt) = 2.303\, \Phi_\lambda I_{0\lambda}\left(\frac{\text{area}}{\text{volume}}\right)\varepsilon(\lambda)Cl \qquad (8)$$

In this case, the reaction rate is directly proportional to the concentration of C. Consequently, the amount of C will decrease exponentially, in the case of photolysis, with time and the reaction will appear to follow first-order kinetics. If a known, constant light intensity is used and the other parameters are known, (dC/dt) can be experimentally determined from the negative slope of a first-order plot of ln C versus exposure time. Thus, the quantum yield can be calculated by rearrangement of equation (8).

When a system is optically thick (i.e. $[\alpha(\lambda) + \varepsilon(\lambda)C]l > 2$) and $\varepsilon(\lambda)C \gg \alpha(\lambda)$, then essentially all of the light incident on the system is absorbed by the photoreactive compound and equation (6) can be written as:

$$-(dC/dt) = \Phi_{a\lambda}I_{0\lambda}\left(\frac{\text{area}}{\text{volume}}\right) \qquad (9)$$

The reaction rate is now independent of concentration. Using a constant light source and sufficiently high concentrations of C in a well-mixed solution, the amount of C will now decrease linearly with time and the reaction will appear to be zero order. Quantum yields can again be determined through rearrangement of equation (9) by using the zero-order rate constant and the other measured parameters.

While direct photolysis is the most straightforward and conceptually simple example of a photochemical process, many marine photochemical transformations are thought to involve photosensitised or indirect secondary pathways.[13-15] In seawater, most secondary photochemical reactions will be initiated by CDOM, although nitrate, nitrite and transition metals may also play a minor role in some waters.[36,37]

Following Hoigné et al.[15] and assuming that CDOM initiates the reaction, the photochemical production rate of a reactive oxidant (Ox) is given by:

$$(d[Ox]/dt)_\lambda = \Phi_{a\lambda,\text{CDOM}}I_{a\lambda,\text{CDOM}} \qquad (10)$$

where [Ox] is in mol L^{-1}, $\Phi_{a\lambda,\text{CDOM}}$ is the apparent quantum yield based on the total photons absorbed by CDOM and is only constant provided that the relative contribution of each decay pathway is invariant with time, and $I_{a\lambda,\text{CDOM}}$ is the specific light absorption rate of CDOM with $a_{\text{CDOM}\lambda}$ (equation 4) substituted for $\varepsilon(\lambda)C$ in equation (7) [note that this changes equation (7) from base 10 to base e and requires $\alpha(\lambda)$ to be converted to base e]. Integration over wavelength yields the total photochemical production rate (r_p).

Various processes in the water column subsequently consume the reactive species produced through the photochemical reaction. The rate of consumption

---

[36] O. C. Zafiriou and M. B. True, *Mar. Chem.*, 1979, **8**, 33.
[37] O. C. Zafiriou and M. B. True, *Geophys. Res. Lett.*, 1979, **6**, 81.

(r_c), which is light independent (if the species is not photoreactive), can be described by a summation of all pseudo-first-order consumption rate process:

$$-(d[\text{Ox}]/dt) = \sum_c (k_{ox,c})[\text{Ox}] \tag{11}$$

When using a constant light source and holding all reaction conditions constant, the system will eventually attain a steady state such that $r_p = r_c$ and a steady-state concentration $[\text{Ox}]_{ss}$ will be reached:

$$[\text{Ox}]_{ss} = \frac{\sum \Phi_{a\lambda,\text{CDOM}} I_{a\lambda,\text{CDOM}}}{\sum (k_{ox,c})} \tag{12}$$

Subsequent *indirect* photochemical reactions that do not significantly alter $[\text{Ox}]_{ss}$ can now be described as pseudo-first-order reaction with the reactive transient:

$$-(dC/dt)_{ox} = k'_{ox}[\text{Ox}]_{ss}C = k_{ox}C \tag{13}$$

where k'_{ox} is the second-order rate constant and k_{ox} is the pseudo-first-order rate constant.

The above discussion highlights the need for careful consideration of light geometry, spectral characteristics and light attenuation in photochemical experiments.[11] Chemical actinometers, compounds with well-established quantum yields and absorption spectra, are extremely useful in determining light intensities within photochemical reaction cells. Several classic chemical actinometers are commonly used[38] and the basic criteria when considering actinometers for use in sunlight irradiations have been described.[39] Proper use of actinometers in the same type of reaction cells and exposed to the same light source, simulated or sunlight, insures that any attenuation, reflection or focusing of light owing to container walls is accounted for in the light measurements. Actinometers may also be used to provide an accurate determination of pathlength within reaction cells.[40]

Considerations of Experimental Design and Interpretation

While experiments with model systems in the laboratory are invaluable for elucidating photochemical reaction methods and rates, they must eventually be compared to field observations to have environmental relevance. For marine photochemists, this entails a number of daunting obstacles. Clearly, the first difficulty is the variability and complexity of CDOM, which acts as the primary light-gathering species for short wavelengths of solar radiation. Its poorly characterised nature prevents the establishment of a direct molar relationship between the primary absorption step and the final photochemical reaction. A second and possibly equally significant problem is the accurate extrapolation of laboratory light data to *in situ* underwater light fields. Even when controlled

---

[38] A. Leifer, *The Kinetics of Environmental Aquatic Photochemistry: Theory and Practice*, American Chemical Society, Washington, 1988.
[39] D. Dulin and T. Mill, *Environ. Sci. Technol.*, 1982, **16**, 815.
[40] R. G. Zepp, *Environ. Sci. Technol.*, 1978, **12**, 327.

irradiations are performed using sunlight, the photochemical rate data are only directly relevant to the upper few centimetres of the water column; that is, provided that any container-wall effects on the light field have been correctly evaluated.[11]

Underwater light fields depend on a wide array of factors including sun position (*i.e.* season, time of day, zenith angle), atmospheric conditions (*i.e.* ozone, clouds, particles), sea surface conditions, concentration of dissolved and suspended material, and depth. Incident solar radiation has been satisfactorily modelled with various degrees of complexity.[10,41–43] Nonetheless, determination of accurate photochemical reaction rates in sunlight requires careful consideration of light intensity and spectral distribution *within* the reaction cells (*vide supra*). The intensity and spectral distribution of incident solar radiation constantly changes, as does the ratio of diffuse to direct radiation. Whereas the former directly influences the underwater light field, the latter affects angles of incidence and consequently light pathlengths in the water column.[10] Excluding the influences of incident radiation, attenuation in the water column will also produce changes in intensity and spectral distribution with depth. Generally speaking, radiative transfer in the water column attenuates the extreme ends of the spectrum most efficiently, producing a spectrum increasingly dominated by blue-green light with depth. Obviously, the exact nature of the spectral changes will vary from site to site. Several detailed treatments of radiative transfer in the water column are available.[24,44,45]

To avoid some of the problems associated with light attenuation in the water column, Zafiriou[3] suggested the use of a 'black sea' model for general approximations. The justification of its use is that all incident solar radiation (except for 5–10% loss due to reflection and backscatter) is absorbed somewhere in the ocean. Consequently, underwater pathlengths are eliminated from consideration. If the diffuse attenuation coefficient ($K_{d\lambda}$, cm^{-1}) of the water column is defined as:[24]

$$K_{d\lambda} = \frac{\ln[I_{d\lambda-0}/I_{d\lambda z}]}{\Delta z} \text{ or } I_{d\lambda z} = I_{d\lambda-0} e^{-K_{d\lambda} z} \tag{14}$$

where $I_{d\lambda-0}$ is downwelling irradiance just below the surface (to negate any reflectance), $I_{d\lambda z}$ is downwelling irradiance at depth z, and z is depth (cm). The vertical extent (cm) of light penetration is approximated by the depth where downwelling irradiance is 1% of $I_{d\lambda-0}$ so that $z(1\%) = 4.6/K_{d\lambda}$. Vertically integrated photochemical production can then be determined by:

$$Pp_\lambda(z) = \int I_{d-0\lambda} e^{-K_{d\lambda} z} \Phi_{a\lambda} a_{CDOM\lambda} d\lambda$$

Integration over z yields:

$$\int \frac{I_{d-0\lambda} \Phi_{a\lambda} a_{CDOM\lambda}}{K_{d\lambda}} d\lambda \tag{15}$$

[41] W. W. Gregg and K. L. Carder, *Limnol. Oceanogr.*, 1990, **35**, 1657.
[42] K. Stamnes, S. C. Tsay, W. Wiscombe and K. Jayaweera, *Appl. Opt.*, 1988, **27**, 2502.
[43] P. Ricchiazzi, S. Yang, C. Gautier and D. Sowle, *Bull. Am. Meteorol. Soc.*, 1998, **79**, 2101.
[44] R. Preisendorf, *Hydrological Optics*, US Dept of Commerce NOAA, Honolulu, 1976.
[45] C. Mobley, *Light and Water*, Academic Press, New York, 1994.

The most obvious problem with this model is when $z_{1\%} > z_{mix}$, as exists in regions where stratification limits the extent of vertical mixing. In these instances, the irradiated water column will not be homogenous with respect to the distribution of possible reactants and products. Several studies have detailed the complexity required to model more rigorously the photochemical processes in relation to mixing and water column optical properties.[46–49] Other studies have used modelling to estimate seasonal and latitudinal variations of photochemical processes[50] and to investigate the distribution and air–sea flux of photochemical products.[51,52]

3 Marine Photoreactants, Products and Processes

Despite the inherent difficulties, many advances have been made in the field of marine photochemistry. These advances have firmly established the need to consider photochemical processes in relation to biogeochemical cycles in general and the marine carbon cycle in particular. This section gives an overview of the reactants, products and processes involved in marine photochemistry.

Marine Chromophores

As mentioned earlier, DOM is believed to be the primary chromophore of the dissolved constituents in seawater. Despite its importance as a light gatherer, its structure is enigmatic and the knowledge of its optical characteristics remains rudimentary. As a striking example, it remains uncertain whether the rather ubiquitous shape of CDOM absorption spectra is the simple sum of a large number of chromophores with differing structures or a small number of chromophores of similar structure with charge transfer interactions.[53] However, there are indications that the chromophores responsible for fluorescence in marine CDOM are distinct from terrestrial CDOM.[54,55] This may be related to some difference in structure, such as the higher percentage of aromaticity in terrestrial *versus* marine DOM.[56] Although some structural properties have been empirically related with optical properties of CDOM,[57–59] there have been no

---

[46] J. M. C. Plane, R. G. Zika, R. G. Zepp and L. A. Burns, in *Photochemistry of Environmental Aquatic Systems*, ed. R. G. Zika and W. J. Cooper, American Chemical Society, Washington, 1987, p. 250.
[47] R. J. Sikorski and R. G. Zika, *J. Geophys. Res.*, 1993, **98**, 2329.
[48] R. J. Sikorski and R. G. Zika, *J. Geophys. Res.*, 1993, **98**, 2315.
[49] S. C. Doney, R. G. Najjar and S. Stewart, *J. Mar. Res.*, 1995, **53**, 341.
[50] A. Kouassi, R. Zika and J. Plane, *Neth. J. Sea Res.*, 1990, **27**, 33.
[51] R. G. Najjar, D. J. Erickson III and S. Madronich, in *The Role of Nonliving Organic Matter in the Earths Carbon Cycle*, ed. R. G. Zepp and C. Sonntag, Wiley, New York, 1995, p. 107.
[52] A. Gnanadesikan, *J. Geophys. Res.*, 1996, **101**, 12.
[53] N. V. Blough and S. A. Green, in *The Role of Nonliving Organic Matter in the Earths Carbon Cycle*, ed. R. G. Zepp and C. Sonntag, Wiley, New York, 1995, p. 23.
[54] P. G. Coble, *Mar. Chem.*, 1996, **51**, 325.
[55] M. M. de Souza Sierra, O. F. X. Donard and M. Lamotte, *Mar. Chem.*, 1997, **58**, 51.
[56] R. Malcolm, *Anal. Chim. Acta*, 1990, **232**, 19.
[57] S. Traina, J. Novak and N. Smeck, *J. Environ. Qual.*, 1990, **19**, 151.
[58] Y.-P. Chin, G. Alken and E. OLoughlin, *Environ. Sci. Technol.*, 1994, **28**, 1853.
[59] C. R. Everett, Y. P. Chin and G. R. Aiken, *Limnol. Oceanogr.*, 1999, **44**, 1316.

systematic efforts to relate optical data to marine DOM chemical properties.[53] Conversely, the few known naturally occurring compounds with significant absorption in the solar spectrum, such as methyl iodide, unsaturated fatty acids, riboflavin, tryptophan, thiamine and vitamin B12, make up a relatively small part of the standing DOM pool.[6] Of course, photochemical transformations may in part be responsible for their low concentrations. Additionally, the contribution to absorption by inorganic chromophores, with the exceptions of nitrite and nitrate, in marine environments is generally a rather small percentage. Suspended particles, both living and non-living, may notably contribute to light absorption, especially in turbid or productive coastal waters.[24,31] Organic compounds adsorbed onto particles display absorption spectra similar to CDOM[60] and have been shown to be photoreactive;[61] however, their impact on marine photochemical processes has been little studied. Detrital particulate material derived from phytoplankton constituents (*i.e.* chlorophyll, pheopigments, carotenoids) have distinctive absorption spectra with characteristic maxima.[62] Photodegradation of these particles yields particles that retain significant absorption of short wavelengths.[62] Many iron species absorb at wavelengths above 300 nm. Despite comprising only a small portion of total absorbance, heterogeneous photochemical reactions at the surface of metal oxides are likely to be important in iron and manganese cycling.[63-66]

Production of Radicals

For many marine waters with CDOM acting as the principal light gatherer, approximately 95–98% of the energy absorbed will be dissipated through decay pathways that do not effect a photochemical change in the system.[16] A small fraction of excited ^{1}CDOM* (1–3%)[67] will produce the relatively long-lived ($> 10^{-3}$ s) triplet species. Owing to the ubiquitous presence of dissolved oxygen, which is an unusual ground state triplet, in surface waters, there is an efficient energy transfer from ^{3}CDOM* to ^{3}O$_2$.[68] This transfer readily yields singlet oxygen ^{1}O$_2$* at 25–100% efficiency due to the lack of spin restrictions, the relatively low activation energy of ^{1}O$_2$* (92 kJ mol^{-1} above the ground state) and the high energy of ^{3}CDOM* (50% are > 250 kJ mol^{-1}).[16] Singlet oxygen does little to promote further photochemical change in seawater owing to either its rapid decay to the ground state or collisional deactivation with water. Its lifetime

---

[60] C. S. Yentsch, *Limnol. Oceanogr.*, 1962, **7**, 207.
[61] G. C. Miller and R. G. Zepp, *Environ. Sci. Technol.*, 1979, **13**, 860.
[62] J. R. Nelson, *Sunlight-Dependent Changes in the Pigment Content and Spectral Characteristics of Particulate Organic Material Derived from Phytoplankton*, Woods Hole Oceanographic Institute Technical Report WHOI-90-09, Woods Hole, 1990.
[63] W. G. Sunda, S. A. Huntsman and G. R. Harvey, *Nature*, 1983, **301**, 234.
[64] W. Stumm and B. Sulzberger, *Geochim. Cosmochim. Acta*, 1992, **56**, 3233.
[65] B. C. Faust, in *Aquatic and Surface Photochemistry*, ed. G. R. Helz, R. G. Zepp and D. G. Crosby, Lewis, Boca Raton, 1994, p. 3.
[66] T. D. Waite and R. Szymczak, in *Aquatic and Surface Photochemistry*, ed. G. R. Helz, R. G. Zepp and D. G. Crosby, Lewis, Boca Raton, 1994, p. 39.
[67] W. R. Haag and T. Mill, *Survey of Sunlight-Produced Transient Reactants in Surface Waters*, Woods Hole Oceanographic Institute Technical Report WHOI-90-09, Woods Hole, 1990.
[68] R. G. Zepp, P. F. Schlotzhauer and R. M. Sink, *Environ. Sci. Technol.*, 1985, **19**, 74.

is of the order of microseconds in the aqueous phase, resulting in low steady-state concentrations[69] and limiting diffusion to < 200 nm.[70,71] Additionally, it is a rather selective oxidant due to its electrophilic nature, but has been suggested to react with electron-rich centres such as dimethyl sulfide.[72,73]

From an effects point of view, the other decay pathways of excited CDOM are the most important. These pathways (right side of Figure 3) produce a variety of both radical and non-radical products (Tables 1 and 2). The exact reaction pathways for most observed products remain currently unresolved. However, many of the primary radicals formed in surface waters will probably react with dissolved oxygen to produce an array of secondary oxygenated radicals.[74] Reported photochemical production rates of radicals in seawater vary greatly, but have been shown to be proportional to the initial CDOM absorbance of the samples. This is not surprising considering equation (6) and the proportional relationship between photochemical rates and absorbance rates in optically thin solutions. As such, and owing to the rarity of quantum yield determinations, photochemical production rates normalised to sample absorbance at 300 or 350 nm can be used to facilitate comparisons of samples from different areas.[17]

Owing to their high reactivity and consequently low concentrations, the measurement of photochemically produced radical species remains problematic[16,74] and thus some species such as CDOM cation radicals, carbonate radicals and dibromide radicals have yet to be rigorously studied. Only hydrogen peroxide accumulates to appreciable concentrations (up to 1000 nmol L^{-1}) in seawater;[75] however, its decay path is principally biologically mediated, which complicates its use as a proxy for strictly photochemical processes.[75,76] The primary formation pathway for hydrogen peroxide is the dismutation of the one-electron reduction product superoxide anion (O_2^-).[16,67] Studies covering a wide range of coastal and open ocean sites have suggested that the superoxide anion is the major photochemically produced radical species in seawater.[77-79] It may be formed from the reduction of dissolved oxygen via either direct electron transfer from excited CDOM or reaction with a hydrated electron.[16] By inference, the production of CDOM cation radicals is expected to be of the same order as superoxide anion. The addition of dioxygen to CDOM carbon-centred radicals is presumed to be the principal source of peroxy radicals in seawater.[16] Excited CDOM is also the dominant source for the generation of the highly reactive

---

[69] W. Haag and J. Hoigné, *Environ. Sci. Technol.*, 1986, **20**, 341.
[70] T. A. Dahl, *Photosensitization and Singlet Oxygen Toxicity*, Woods Hole Oceanographic Institute Technical Report WHOI-90-09, Woods Hole, 1990.
[71] T. A. Dahl, in *Aquatic and Surface Photochemistry*, ed. G. R. Helz, R. G. Zepp and D. G. Crosby, Lewis, Boca Raton, 1994, p. 241.
[72] P. Brimblecombe and D. Shooter, *Mar. Chem.*, 1986, **19**, 343.
[73] D. Kieber, J. Jiao, R. Kiene and T. Bates, *J. Geophys. Res.*, 1996, **101**, 3715.
[74] O. C. Zafiriou, N. V. Blough, E. Micinski, B. Dister, D. Kieber and J. Moffett, *Mar. Chem.*, 1990, **30**, 45.
[75] R. G. Zika, *Hydrogen Peroxide as a Relative Indicator of the Photochemical Reactivity of Natural Waters*, Woods Hole Oceanographic Institute Technical Report WHOI-90-09, Woods Hole, 1990.
[76] R. G. Petasne and R. G. Zika, *Mar. Chem.*, 1997, **56**, 215.
[77] O. C. Zafiriou and B. Dister, *J. Geophys. Res.*, 1991, **96**, 4939.
[78] E. Micinski, L. A. Ball and O. C. Zafiriou, *J. Geophys. Res.*, 1993, **98**, 2299.
[79] B. Dister and O. C. Zafiriou, *J. Geophys. Res.*, 1993, **98**, 2341.

Table 1 Photochemically formed radicals in freshwater and seawater[67]

Species	Estimated conc. (mol L^{-1})	Probable pathways	Refs.
CDOM triplet (^{3}CDOM*)	~10^{-10}	CDOM intersystem crossing	68
CDOM cation (CDOM$^+$)	~10^{-10}	CDOM electron ejection and reactions with O_2	80
Singlet oxygen ($^1O_2$*)	10^{-14}–10^{-13}	Energy transfer from ^{3}CDOM*	68, 69
Superoxide anion (O_2^-)	10^{-9}–10^{-8}	e^- transfer to and e^-_{aq} reaction with $^3O_2$	81
Hydroxyl radicals (OH)	10^{-19}–10^{-17}	Nitrate, nitrite and CDOM photolysis	36, 37, 82
Hydrogen peroxide (H_2O_2)	10^{-8}–10^{-7}	Dismutation of O_2	83
Hydrated electron (e^-_{aq})	10^{-17}–10^{-15}	CDOM electron ejection	80
Organoperoxides (ROO)	10^{-11}–10^{-10}	CDOM photolysis	84
Carbon-centred radicals (RH_2C)	10^{-13}–10^{-11}	CDOM photolysis	85
Carbonate radical (CO_3^-)	~10^{-14}	OH/HCO_3^- (FW) and Br_2^-/HCO_3^- (SW) reactions	86
Dibromide anion radical (Br_2^-)	~10^{-14}	OH/Br^- reactions	87, 88
Cu^+, Mn^{2+}, Fe^{2+}	10^{-12}–10^{-9}	Ligand-to-metal charge transfers	66, 89, 90

hydroxyl radical in seawater.[82,87] The production of hydroxyl radicals from CDOM can occur in both aerobic and anaerobic natural waters, possibly through hydrogen atom extraction from water by excited quinone groups within the CDOM structure.[91] An estimated 93% of hydroxyl radicals produced will react with bromide in seawater.[82] The remaining 7% and the possible dibromide anion radical daughter products are free to participate in further reactions with DOM.

[80] R. G. Zepp, A. M. Braun, J. Hoigné and J. A. Leenheer, *Environ. Sci. Technol.*, 1987, **21**, 485.
[81] R. G. Petasne and R. G. Zika, *Nature*, 1987, **325**, 516.
[82] K. Mopper and X. Zhou, *Science*, 1990, **250**, 661.
[83] W. J. Cooper, R. G. Zika, R. G. Petasne and J. M. C. Plane, *Environ. Sci. Technol.*, 1988, **22**, 1156.
[84] B. C. Faust and J. Hoigné, *Environ. Sci. Technol.*, 1987, **21**, 957.
[85] D. J. Kieber and N. V. Blough, *Anal. Chem.*, 1990, **62**, 2275.
[86] R. A. Larson and R. G. Zepp, *Environ. Toxicol. Chem.*, 1988, **7**, 265.
[87] X. Zhou and K. Mopper, *Mar. Chem.*, 1990, **30**, 71.
[88] O. C. Zafiriou, M. B. True and E. Hayon, in *Photochemistry of Environmental Aquatic Systems*, ed. R. G. Zika and W. J. Cooper, American Chemical Society, Washington, 1987, p. 89.
[89] J. Sykora, *Coord. Chem. Rev.*, 1997, **159**, 95.
[90] D. W. King, H. A. Lounsbury and F. J. Millero, *Environ. Sci. Technol.*, 1995, **29**, 818.
[91] P. P. Vaughan and N. V. Blough, *Environ. Sci. Technol.*, 1998, **32**, 2947.

Table 2 Identified products from photochemical reactions involving DOM[17]

Compound	Refs.
Acetaldehyde	92–94
Acetate	95–97
Acetone	93
Citrate	95
Formaldehyde	92, 94
Formate	95–97
Glyoxal	93–94
Glyoxyalate	92, 94, 98
Levulinate	95
Malonate	96, 97
Methylglyoxal	93
Oxalate	96, 97
Propanal	93
Pyruvate	94, 95, 98, 99
Non-methane hydrocarbons	100, 101
Carbon monoxide	102–105
Carbon dioxide	103, 106
Carbonyl sulfide	107–111
Carbon disulfide	112
Dimethyl sulfoxide	73
Methyl iodide	113
Phosphate	114
Nitrite	115, 116
Ammonium	103, 117

[92] R. J. Kieber, X. Zhou and K. Mopper, *Limnol. Oceanogr.*, 1990, **35**, 1503.
[93] K. Mopper and W. L. Stahovec, *Mar. Chem.*, 1986, **19**, 305.
[94] K. Mopper, X. Zhou, R. J. Kieber, D. J. Kieber, R. J. Sikorski and R. D. Jones, *Nature*, 1991, **353**, 60.
[95] R. G. Wetzel, P. G. Hatcher and T. S. Bianchi, *Limnol. Oceanogr.*, 1995, **40**, 1369.
[96] S. Bertilsson and L. Tranvik, *Limnol. Oceanogr.*, 1998, **43**, 885.
[97] J. Dahlén, S. Bertilsson and C. Pettersson, *Environ. Int.*, 1996, **22**, 501.
[98] D. J. Kieber and K. Mopper, *Mar. Chem.*, 1987, **21**, 135.
[99] D. J. Kieber, J. A. McDaniel and K. Mopper, *Nature*, 1989, **341**, 637.
[100] D. D. Riemer, W. H. Pos and R. G. Zika, Program and Abstracts from the 1997 ASLO Ocean Sciences Meeting, Sante Fe, NM, 1997, p. 282.
[101] A. C. Lewis, J. B. McQuaid, N. Carslaw and M. J. Pilling, *Atmos. Environ.*, 1999, **33**, 2417.
[102] R. L. Valentine and R. G. Zepp, *Environ. Sci. Technol.*, 1993, **27**, 409.
[103] H. Gao and R. G. Zepp, *Environ. Sci. Technol.*, 1998, **32**, 2940.
[104] R. D. Jones, *Deep-Sea Res.*, 1991, **38**, 625.
[105] Y. Zuo and R. D. Jones, *Naturwissenschaften*, 1995, **82**, 472.
[106] W. L. Miller and R. G. Zepp, *Geophys. Res. Lett.*, 1995, **22**, 417.
[107] M. O. Andreae and R. J. Ferek, *Glob. Biogeochem. Cycl.*, 1992, **6**, 175.
[108] R. G. Zepp and M. O. Andreae, *Geophys. Res. Lett.*, 1994, **21**, 2813.
[109] P. S. Weiss, S. S. Andrews, J. E. Johnson and O. C. Zafiriou, *Geophys. Res. Lett.*, 1995, **22**, 215.
[110] G. Uher and M. O. Andreae, *Limnol. Oceanogr.*, 1997, **42**, 432.
[111] O. Flöck and M. Andreae, *Mar. Chem.*, 1996, **54**, 11.
[112] H. Xie, R. Moore and W. Miller, *J. Geophys. Res.*, 1998, **103**, 5635.
[113] R. Moore and O. C. Zafiriou, *J. Geophys. Res.*, 1994, **99**, 16.
[114] D. A. Francko and R. T. Heath, *Limnol. Oceanogr.*, 1982, **27**, 564.
[115] R. Kieber, A. Li and P. Seaton, *Environ. Sci. Technol.*, 1999, **33**, 993.
[116] L. Spokes and P. Liss, *Mar. Chem.*, 1996, **54**, 1.
[117] K. L. Bushaw, R. G. Zepp, M. A. Tarr, D. Schulz-Jander, R. A. Bourbonniere, D. A. Bronk and M. A. Moran, *Nature*, 1996, **381**, 404.

Production of Stable Products

The cascade of radical species resulting from the photolysis of DOM in seawater ultimately terminates in the production of stable non-radical species. Although the exact decay pathways are not known, a sizeable number of low molecular weight (LMW) compounds (Table 2) have been measured as the result of DOM photolysis. These are most likely produced from radical intermediates and fragmentation reactions arising from the net oxidative flow of electrons from CDOM to dioxygen.[118] However, some LMW products may not be structurally part of the uncharacterised DOM fraction, but merely adsorbed onto humic DOM and released upon irradiation.[119,120] Carbonyl moieties within the DOM structure may be particularly prone to photochemical reactions.[121] Some researchers have suggested that the direct production of CO[102,122] and CO_2[103,106] upon DOM photolysis arises through decarbonylation reactions.[123] Carbon monoxide and carbonyl sulfide (OCS) are thought to be photochemically produced via a common intermediate derived from a carbonyl moiety in DOM.[124] Carbonyl groups may also be involved in reactions producing non-methane hydrocarbons (NMHC) in seawater.[100] Regardless of the exact pathways involved, the production of LMW compounds has important consequences for carbon cycling in the oceans (*vide infra*).

Photochemical reactions produce volatile species that can subsequently be involved in exchange with the atmosphere. Besides CO and CO_2, other gaseous species identified as photochemical products include carbonyl sulfide,[107–110, 124] carbon disulfide (CS_2),[112] methyl iodide (CH_3I)[113] and NMHC.[100,107,114] Emission of these photoproducts to the atmosphere may subsequently influence atmospheric chemistry. For example, up to 88% of NMHC reactions with atmospheric hydroxyl radicals at a coastal marine site involved NMHC derived from ocean photochemical processes.[107] Likewise, the enhanced sea-to-air emission of reduced sulfur species due to marine photochemical production and subsequent production of sulfate aerosols may the influence atmospheric radiation budget and heterogeneous chemical processes in the atmosphere such as ozone destruction.[19] Conversely, photochemical oxidation may limit the ocean–atmosphere transfer of dimethyl sulfide (DMS), another important contributor to atmospheric sulfate aerosols, through the production of highly soluble dimethyl sulfoxide (DMSO) or other non-volatile sulfur products.[72,73] Although DMS photolysis can be photosensitised by DOM,[126] the only published quantum yield spectrum for DMS photolysis is unusual in that a distinct maximum is observed around 400 nm.[73] Thus, the DMS quantum yield

---

[118] N. V. Blough, in *The Sea Surface and Global Change*, ed. P. Liss and R. Duce, Cambridge University Press, Cambridge, 1997, p. 383.
[119] J. A. Amador, M. Alexander and R. G. Zika, *Appl. Environ. Microbiol.*, 1989, **55**, 2843.
[120] J. A. Amador, M. Alexander and R. G. Zika, *Environ. Toxicol. Chem.*, 1991, **10**, 475.
[121] J. Kagan, *Organic Photochemistry, Principles and Applications*, Academic Press, San Diego, 1993.
[122] Y. Zuo and R. D. Jones, *Water Res.*, 1997, **31**, 850.
[123] Y. Chen, S. U. Khan and M. Schnitzer, *Soil Sci. Soc. Am. J.*, 1978, **42**, 292.
[124] W. Pos, D. Riemer and R. Zika, *Mar. Chem.*, 1998, **62**, 89.
[125] R. F. Lee and J. Baker, *Mar. Chem.*, 1992, **38**, 25.
[126] A. Brugger, D. Slezak, I. Obernosterer and G. Herndl, *Mar. Chem.*, 1998, **59**, 321.

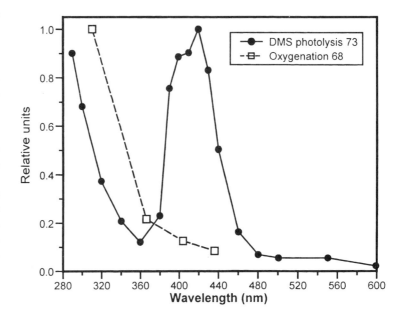

Figure 5 A comparison of the relative quantum yield spectrum for DMS photolysis and photo-oxygenation attributable to singlet oxygen. References are given for each process. The spectra are normalised to 1 at the wavelength of the maximum rate. The spectrum for DMS is currently the only one published for aquatic photoprocesses that has a maximum not located in the UV region

spectrum is in stark contrast to DOM photosensitised oxygenation spectra[68] (Figure 5) and suggests a relatively specific, but as yet unidentified, photolysis pathway.

Gaseous and volatile species may be more prone to sea–air exchange when their production occurs in the sea surface microlayer. The sea surface microlayer is enriched in DOM, CDOM, trace metals, particulate matter and microorgansims as compared to bulk seawater.[118] Measured photochemical production rates of LMW products are 1.2 to 25 (composite average 4.3) fold faster in the microlayer, which is roughly equivalent to the enrichment in concentrations compared to the underlying waters.[127] Modelling efforts indicate that, depending on atmospheric concentrations and partitioning coefficients between gaseous and aqueous phases, photochemical production in the microlayer may be sufficient to drive a net sea–air exchange for some LMW products.[127] However, classical exchange models were not consistent with concentration data and thus the role of microlayer photochemistry in sea–air exchange remains ill-defined.[127]

Photochemical production of inorganic species has also been shown. The release of complexed/bound phosphate from UV irradiated humic material has been shown in freshwater;[114] however, a corresponding process in seawater has not yet been observed. Photochemical production of ammonium from a variety of freshwater samples and one estuarine sample has been reported,[117] as well as from isolated estuarine humic material.[128] Photochemical ammonification has been estimated to increase the availability of terrestrially derived N by 20% on the south-eastern continental shelf of the USA.[117] Recently, the photoproduction of nitrite from DOM in freshwater and seawater was demonstrated.[115] Nitrite photoproduction competes with the direct photolysis of nitrite,[37] but photolysis

[127] X. Zhou and K. Mopper, *Mar. Chem.*, 1997, **56**, 201.
[128] K. L. Bushaw-Newton and M. A. Moran, *Aquat. Microb. Ecol.*, 1999, **18**, 285.

is about six times faster than production. As such, net photochemical production or loss would depend on the relative concentrations of nitrite and DOM in the water sample.[115] Other researchers have arrived at a similar conclusion,[116] but suggested that some of the nitrite formed may come from the photosensitised or direct reduction of nitrate.[36]

Photochemical processes also play an important role in redox reactions involving trace metals in seawater. Iron has been suggested to be a limiting micro-nutrient for primary production in certain ocean areas.[129] Although the exact speciation of inorganic iron in seawater remains equivocal,[65] iron(III) (hydr)oxides are the most thermodynamically stable species in oxygenated seawater. However, recent evidence indicates that organic iron complexes may dominate iron speciation in seawater.[130–131] In either case, numerous studies have shown that iron can be photoreduced[65] and measurable concentrations of dissolved iron occur in irradiated, oxygenated seawater.[132] Copper, another essential trace element, is present mainly as Cu(II) organic complexes in seawater. Some of these complexes are photoreactive,[118] which may in part explain surface maxima in vertical profiles of Cu(I).[16] Although colloidal Mn(IV) oxides are thermodynamically favoured, manganese as Mn(II) dominates in surface layers of the sea.[16] In the absence or photoinhibition of certain bacteria, Mn(II) is kinetically stable to oxidation in seawater. Thus, surface maxima of Mn(II) are attributable to the combined effects of photoreduction of manganese oxides and photoinhibition of oxidising bacteria.[133]

Photobleaching and Phototransformations of CDOM

In addition to the production of identified LMW products, sunlight exposure causes the loss or photobleaching of CDOM absorbance properties.[25,109,120,134–136] In filtered seawater samples, the production rate of identifiable compounds was less than 20% of the photobleaching rate,[113] indicating that the majority of photoproducts are yet to be identified. Although little is known of the exact nature of the photobleached products, they are likely to be LMW compounds resulting from the cleavage of high molecular weight DOM.[122,137,138] These uncharacterised LMW products, along with identified products (Table 2), may contribute to increases in bacterial consumption of otherwise biologically refractory DOM in post-irradiation incubations.[139,140] Conversely, decreases in bacterial activity observed in post-irradiation incubations

---

[129] J. H. Martin and S. E. Fitzwater, *Nature*, 1988, **331**, 341.
[130] E. L. Rue and K. W. Bruland, *Mar. Chem.*, 1995, **50**, 117.
[131] C. M. G. van den Berg, *Mar. Chem.*, 1995, **50**, 139.
[132] W. L. Miller, D. W. King, J. Lin and D. A. Kester, *Mar. Chem.*, 1995, **50**, 63.
[133] W. G. Sunda and S. A. Huntsman, *Limnol. Oceanogr.*, 1990, **35**, 325.
[134] A. M. Kouassi and R. G. Zika, *Toxicol. Environ. Chem.*, 1992, **35**, 195.
[135] I. Reche, M. L. Pace and J. J. Cole, *Biogeochemistry*, 1999, **44**, 259.
[136] R. F. Whitehead, S. J. de Mora, S. Demers, P. Monfort and B. Mostajir, *Limnol. Oceanogr.*, 2000, **45**, 278.
[137] H. De Haan, *Limnol. Oceanogr.*, 1993, **38**, 1072.
[138] D. Hongve, *Acta Hydrochim. Hydrobiol.*, 1994, **22**, 117.
[139] W. L. Miller and M. A. Moran, *Limnol. Oceanogr.*, 1997, **42**, 1317.
[140] I. Reche, M. L. Pace and J. J. Cole, *Microb. Ecol.*, 1998, **36**, 270.

of waters with freshly produced algal DOM[141,142] may indicate diminished substrate bioavailability.[143] Photo-induced polymerisation of labile algal exudates, such as polyunsaturated fatty acids (PUFAs), have been suggested as one pathway for the production of marine humic materials[144–146] and the process is associated with an increase in absorbance after irradiation.[147] This process is not likely to be seen in most filtered water studies owing to the low concentrations and rapid turnover of PUFAs or other labile exudates. However, a continual re-supply of freshly produced DOM from concurrent planktonic activity may affect CDOM absorbance and photobleaching rates *in situ*.[136,148]

4 UV-B Radiation and Global Significance for Marine Biogeochemical Cycles

Estimates of Global Photochemical Production

A paradox in the ocean carbon cycle is how to reconcile the old apparent age of deep water DOM (4000–6000 yr bp),[149,150] ocean circulation times (~ 1400 yr),[151] the lack of significant terrestrial influence observed in open ocean DOM[30,149] and the substantial inputs of biologically recalcitrant river-transported DOM ($\sim 0.25 \times 10^{15}$ g C yr^{-1}).[152] For example, a recent study showed that only 1–3% of riverine DOM from five estuaries in the USA was consumed by estuarine bacteria.[153] This study is consistent with a more extensive literature review that estimated $19 \pm 16\%$ of riverine DOM was biologically labile.[154] Additionally, carbon isotopic data indicate that estuarine bacteria almost exclusively assimilate modern (*i.e.* autochthonous) DOM.[155] These recent studies support the general view of the biologically refractory nature of riverine DOM[27] and the more or less conservative mixing nature of DOM in estuaries.[156] However, if riverine DOM were entirely refractory, simple calculations show that the amount of DOM delivered to the oceans by rivers is sufficient to replace the entire ocean DOM pool in < 2000 years,[157] a result clearly at odds with field observations of the non-terrestrial characteristics of oceanic DOM (*vide supra*).

[141] C. J. Gobler, D. A. Hutchins, N. S. Fisher, E. M. Cosper and S. A. Sañudo-Wilhelmy, *Limnol. Oceanogr.*, 1997, **42**, 1492.
[142] R. Benner and B. Biddanda, *Limnol. Oceanogr.*, 1998, **43**, 1373.
[143] I. Obernosterer, B. Reitner and G. Herndl, *Limnol. Oceanogr.*, 1999, **44**, 1645.
[144] G. Harvey, D. Boran, L. Chesal and J. Tokar, *Mar. Chem.*, 1983, **12**, 119.
[145] G. Harvey, D. Boran, S. Piotrowicz and C. Weisel, *Nature*, 1984, **309**, 244.
[146] R. Kieber, L. Hydro and P. Seaton, *Limnol. Oceanogr.*, 1997, **42**, 1454.
[147] J. Wheeler, *J. Geophys. Res.*, 1972, **77**, 5302.
[148] M. Vernet and K. Whitehead, *Mar. Biol.*, 1996, **127**, 35.
[149] P. M. Williams and E. R. M. Druffel, *Nature*, 1987, **330**, 246.
[150] J. E. Bauer, P. M. Williams and E. R. M. Druffel, *Nature*, 1992, **357**, 667.
[151] W. Broecker, *Chemical Oceanography*, Harcourt Brace Jovanovich, New York, 1974.
[152] M. Meybeck, *Am. J. Sci.*, 1982, **282**, 401.
[153] M. A. Moran, W. M. Sheldon and J. E. Sheldon, *Estuaries*, 1999, **22**, 55.
[154] M. Søndergaard and M. Middelboe, *Mar. Ecol. Prog. Ser.*, 1995, **118**, 283.
[155] J. Cherrier, J. E. Bauer, E. R. M. Druffel, R. B. Coffin and J. P. Chanton, *Limnol. Oceanogr.*, 1999, **44**, 730.
[156] R. F. C. Mantoura and E. M. S. Woodward, *Geochim. Cosmochim. Acta*, 1983, **47**, 1293.
[157] W. G. Deuser, *Nature*, 1988, **332**, 396.

Part of the puzzle may be explained by an expanding set of observations showing that biological recalcitrance does not equate with chemical inertness. Thus, the photochemical production of LMW compounds (Table 2) or other unidentified compounds that are more readily bioavailable is a mechanism by which chemical processes may 'prime' the bacterial utilisation of recalcitrant DOM.[158] Numerous studies have confirmed that bacterial activity can be enhanced in light-exposed samples as compared to samples held in the dark[17] and especially in samples with strong terrestrial influence. Additionally, the photodegradation of DOM to dissolved inorganic carbon or other volatile compounds creates a removal pathway through sea–air exchange.

Global estimates of marine photochemical processes are limited by the considerable variability of CDOM concentrations and absorption and the lack of sufficient data on differences between marine and terrestrial CDOM photoprocesses. However, some constraints on the importance of photochemical processes in the marine biogeochemical cycle can be made. For example, using the ratio of CO:LMW and CO:ammonium production ratios in conjunction with CO quantum yield data, Moran and Zepp[17] calculated that at least 1.7 Pg C yr^{-1} and 0.15 Pg N yr^{-1} are produced globally in the oceans, assuming that all solar UV radiation is absorbed by CDOM. These values correspond to approximately 4% and 5% of the annual global bacterial carbon and nitrogen demand in the oceans, respectively. If unidentified photobleached products were included, the estimate of biologically available C may be even higher. In addition, the direct production of CO_2 from photodegradation of DOM is estimated to be 12–16 Pg C yr^{-1} globally. The combined estimated carbon fluxes represent $\sim$ 2–3% of the oceanic DOC pool (700 Pg C) and are approximately 80-fold greater than the annual river-carried DOC (0.2 Pg C yr^{-1}) input to the ocean. In comparison with other ocean carbon fluxes, photochemical C fluxes are $\sim$ 30% of the global average for total marine primary production (43.5–46.9 Pg C yr^{-1})[159] and about 150-fold greater than global estimates for organic carbon burial in sediments (0.1 Pg C yr^{-1}).[27] The photochemical C transformation estimates are conservative in that unidentified products are not included. However, because photobleaching rates are generally faster than photoproduction rates, the loss of CDOM absorbance may limit the extent of photochemical production[106] and thereby reduce these estimates. Further research is required to refine understanding of these processes, but the estimates clearly demonstrate a potentially important role for photochemical processes in ocean biogeochemical cycling.

Influence of Ozone Depletion on Photochemical Production

The implicit assumption throughout this discussion has been that marine photochemistry is almost exclusively UV photochemistry. This assumption is based on the ubiquitous shape of CDOM absorption spectra (Figure 2) and the similarity of published quantum yield spectra for a number of marine photochemical processes (Figure 4). In fact, of the quantum yield spectra published to date for marine photochemical processes, only one (for DMS photolysis)[73] shows a

[158] D. J. Strome and M. C. Miller, *Int. Ver. Theor. Angew. Limnol. Verh.*, 1978, **20**, 1248.
[159] M. J. Behrenfield and P. G. Falkowski, *Limnol. Oceanogr.*, 1997, **42**, 1479.

maximum outside the UV-B (280–320) range. However, because photochemical rates also depend on incident solar radiation and attenuation in the water column, maximum rates near the surface often occur in the UV-A (320–400 nm) range and may be found extending into the visible (400–700 nm) range at depth.[17] Although increases in UV-B radiation due to ozone depletion[22] will undoubtedly affect photochemical rates, the question of the significance of the effect remains ill-defined. The global distribution of ozone depletion suggests that changes in UV-B will be greatest at high latitudes whereas equatorial and tropical latitudes will be least affected. On the other hand, global distributions of photochemical production[17,108] indicate that maximum production occurs at low latitudes owing to higher average solar irradiance.

As a first approximation of the influence of ozone depletion on marine photochemical processes, the 1979–1999 global ozone record from the Total Ozone Mapping Spectrometer (TOMS) satellite data can be used to determine zonal, monthly trends in column ozone. To estimate monthly photochemical production over 10° latitude bands, the following procedure is followed. A cloudless, typical marine atmosphere is used with a standard vertical ozone profile[160] scaled to the monthly mean total column ozone from TOMS (Nimbus 7, Meteor 3, Earth Probe; ver 7) satellite observations for each latitude band. The atmospheric profile is then used by a discreet ordinate radiative transfer model[43] to generate the downwelling spectral irradiance (I_0) at 1 nm intervals over the 280–650 nm wavelength interval. The albedo is assumed to be isotropic and set to 10%. The process is repeated at 1-h increments for the 15th day of each of the 216 months of the TOMS ozone record. At each time point, vertically integrated photochemical production is calculated using equation (15). To facilitate comparisons with previous estimates (*vide supra*), the following assumptions are used in equation (15): a_{CDOM} of 5.5 m^{-1} and 2.7 m^{-1} at 300 and 350 nm, respectively, and an a_{CDOM} absorption spectrum based on $a_\lambda = a_{300} e^{-0.014(300-\lambda)}$;[25] $K_{d\lambda}$ calculated from the bio-optical model of Smith and Baker[161] with chlorophyll at 1 mg m^{-3}; and the quantum yield data for CO and formaldehyde from Figure 4. The results are integrated over time to estimate a mean monthly, vertically integrated photochemical production rate. Trends in the monthly means for each latitude band are determined by using the slopes of the least squares regression lines over the data record to calculate the percent change per decade relative to 1979 (Figure 6). As expected based on the ozone trends, the largest increases in photochemical production rates occur in the high latitudes of the Southern Hemisphere during the spring with a corresponding, but much weaker, peak in the Northern Hemisphere. The equatorial and lower tropical latitudes, where ozone changes are not significantly different from zero at the 95% confidence interval, have statistically insignificant trends throughout the year. The striking difference in the magnitude of the increases of formaldehyde production (~19% per decade maximum increase) and carbon monoxide (~3% per decade maximum increase) illustrates the need for well-defined quantum yield

[160] R. A. McClatchey, R. W. Fenn, J. E. A. Selby, F. E. Volz and J.S. Garing, *Optical Properties of the Atmosphere*, 3rd edn., Air Force Cambridge Research Laboratories Report AFCRL-72-0497, Bedford, MA, 1972.
[161] R. C. Smith and K. S. Baker, *Limnol. Oceanogr.*, 1982, **27**, 500.

Figure 6 Monthly trends, given as % increase per decade, for the photoproduction of (a) carbon monoxide and (b) formaldehyde. As described in the text, these values were derived using modelled increases in UV-B radiation based on TOMS satellite observations of decreases in total column ozone over the period from 1979 to 1999. The hatched areas show where the trends in ozone changes are not significantly different from zero at the 95% confidence level

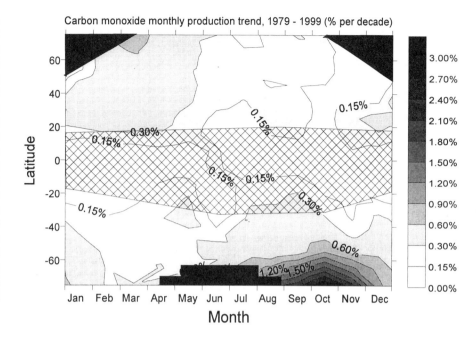

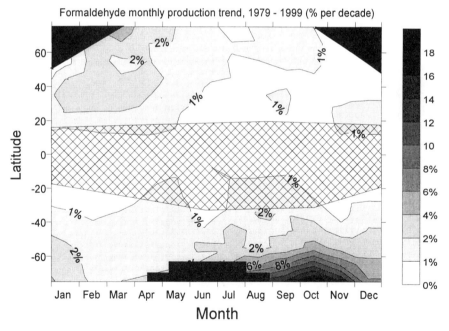

data. Whereas the published quantum yield data for formaldehyde is confined to the UV-B region, the exponential tail of the CO quantum yield spectrum extends well into the visible range (Figure 4). This difference strongly influences the response to changes in UV-B due to ozone depletion. In the case of CO, the response to changes in UV-B is measured against a background production from longer wavelengths and thus the relative change is small. Formaldehyde production, on the other hand, depends entirely on UV-B wavelengths and thus is much more sensitive to changes in ozone concentrations.

5 Summary and Conclusions

The absorption of sunlight by seawater constituents, chiefly ill-defined dissolved organic compounds, triggers a cascade of reactions producing a wide array of radical and non-radical species. A large percentage of the final stable products remain as enigmatic as the exact pathways and reactants that produce them. Equally daunting are the experimental considerations required to apply laboratory results in the prediction of *in situ* photochemical processes. Nevertheless, understanding of marine photochemistry has advanced to the point where realistic constraints can be placed on its global significance. The research to date indicates that photochemical processes may play a fundamental role in the ocean carbon cycle. The biogeochemical cycling of other elements, such as sulfur, nitrogen, phosphorus and trace metals, may also be affected. The effects on the elements that serve as macro- or micro-nutrients for marine plankton may promote an increase biological production in some areas of the oceans. On the other hand, photochemically produced reactive radical species may serve as an additional stress on microorganisms. The net result of the positive and negative effects of photochemical reactions on marine primary and secondary production has important implications for the functioning of marine ecosystems and their responses to changes in solar radiation entering the ocean. In this context, the consequences of increased UV-B radiation due to ozone depletion on marine ecosystems are likely to involve an interplay of photobiological and photochemical responses. Although research directed specifically at the ramifications of increased UV-B is important, the continued exploration of basic photochemistry in the oceans is vital to broaden the understanding of specific processes and to refine how they can be generalised to photochemical processes on regional and global scales.

Assessing Biological and Chemical Effects of UV in the Marine Environment: Spectral Weighting Functions

PATRICK J. NEALE AND DAVID J. KIEBER

1 Introduction

All matter in surface waters, both living and non-living, is exposed to sunlight and, in particular, solar ultraviolet radiation (UV; 280–400 nm). This UV radiation induces excited states in a wide variety of molecules, with many consequences. For non-living matter, UV exposure induces chemical transformations and changes the chemical profile of natural waters. For organisms in those waters, chemical transformation of basic biomolecules—nucleic acids, proteins, lipids—slows metabolism and can completely stop life or at least reproduction, if the exposure is severe enough. Dissolved or colloidal substances and microorganisms are directly exposed to UV and show the most profound responses to solar exposure. Dissolved organic matter absorbs most of the UV incident on natural waters and participates in most UV-dependent photochemistry (see Chapter 3, this volume). The damaging potential of UV for some biological processes such as phytoplankton photosynthesis has been known for some time,[1] but the scale and scope of investigation into the effects of UV have considerably expanded over the last two decades with the concurrent decline in stratospheric ozone.[2-5] Ozone depletion increases surface incident UV, with the most pronounced changes in the short-wavelength UV-B (280–320 nm).[6] A special

---

[1] E. Steemann Nielsen, *J. Cons. Perm. Int. Explor. Mer*, 1964, **22**, 130.
[2] D. Karentz, M. L. Bothwell, R. B. Coffin, A. Hanson, G. J. Herndl, S. S. Kilham, M. P. Lesser, M. Lindell, R. E. Moeller, D. P. Morris, P. J. Neale, R. W. Sanders, C. S. Weiler, and R. G. Wetzel, *Arch. Hydrobiol. Beih Ergebn. Limnol.*, 1994, **43**, 31.
[3] D.-P. Häder, *The Effects of Ozone Depletion on Aquatic Ecosystems*, Landes, Georgetown, TX, 1997.
[4] S. J. de Mora, S. Demers, and M. Vernet (eds.), *The Effects of UV Radiation on Marine Ecosystems*, Cambridge University Press, Cambridge, 2000.
[5] C. R. Booth, J. H. Morrow, T. P. Coohill, J. E. Frederick, D.-P. Häder, O. Holm-Hansen, W. H. Jeffrey, D. L. Mitchell, P. J. Neale, I. Sobolev, J. van der Leun, and R. C. Worrest, *Photochem. Photobiol.*, 1997, **65**, 252.
[6] A. R. Webb, this volume, Chapter 2.

focus has been given to the effects of UV on biological and chemical processes in the Southern Ocean, which lies under the seasonal Antarctic 'ozone hole'.[7]

As documented in the previous chapter, almost all marine photochemistry can be attributed to the effects of UV and short-wavelength visible light. In the Southern Ocean and elsewhere, it has been shown that most microbial life is sensitive to *in situ* UV exposure, including viruses, bacteria, phytoplankton, zooplankton, and larval forms of many organisms.[2-4] However, much of the early research is qualitative, and it is only recently that progress has been made in quantifying the biogeochemical significance of UV effects on marine processes, as well as the particular effect of increased UV-B due to ozone depletion. In this chapter we review efforts directed towards obtaining quantitative estimates of the effect of UV on marine chemistry and biology.

Spectral Dependence of UV Effects

The importance of UV exposure to aquatic processes has motivated the development of models that define how impacts may increase with ozone depletion or be affected by changes in the UV transparency of natural waters (largely a function of chromophoric dissolved organic matter or CDOM).[8,9] These environmental changes bring about wavelength-dependent changes in UV, and in order to understand the biological and chemical effects of those changes, wavelength-dependent weighting functions are required.[10,11] Spectral weighting functions provide an approach to properly scale different exposure spectra for their 'effectiveness' in inducing a chemical reaction with attendant biological consequences. The concept is analogous to computing a weighted average of a UV spectrum, in that the weighting coefficients emphasize the shorter (more damaging) wavelengths relative to the longer wavelengths. The approach is to calculate a single measure of damaging (or photochemically active) irradiance, E^*, by summing the product of spectral exposure, $E(\lambda)$ (mW m^{-2} nm^{-1} or mol photons cm^{-2} s^{-1} nm^{-1}), and a scaling coefficient or weight, $\varepsilon(\lambda)$, over a series of narrow wavelength bands ($\Delta\lambda$):

$$E^* = \sum_{\lambda=280\,nm}^{400\,nm} \varepsilon(\lambda) \cdot E(\lambda) \cdot \Delta\lambda \qquad (1)$$

If the $\varepsilon(\lambda)$ are chosen correctly, then the UV effect will be a single function of E^* no matter what is the spectral composition $E(\lambda)$.

The spectral difference in $\varepsilon(\lambda)$ is fundamentally related to the variation in reaction rate(s) across the UV spectrum. The reaction rate is the product of two

---

[7] C. S. Weiler and P. A. Penhale (eds.), *Ultraviolet Radiation in Antarctica: Measurement and Biological Effects*, American Geophysical Union, Washington, 1994.
[8] P. J. Neale, in *The Effects of UV Radiation on Marine Ecosystems*, ed. S. J. de Mora, S. Demers, and M. Vernet, Cambridge University Press, Cambridge, 2000, p. 73.
[9] J. J. Cullen and P. J. Neale, in *Effects of Ozone Depletion on Aquatic Ecosystems*, ed. D.-P. Häder, Landes, Austin, TX, 1997, p. 97.
[10] T. P. Coohill, *Photochem. Photobiol.*, 1989, **50**, 451.
[11] M. M. Caldwell, L. B. Camp, C. W. Warner, and S. D. Flint, in *Stratospheric Ozone Reduction, Solar Ultraviolet Radiation and Plant Life*, ed. R. C. Worrest and M. M. Caldwell, Springer, Berlin, 1986, p. 87.

probabilities: the probability that an incident photon is absorbed and the probability that an absorbed photon will bring about the reaction. The latter probability is called the photochemical quantum yield.[12] In environmental samples, both the absorbance of CDOM and the quantum yield of most reactions vary with wavelength, with shorter wavelengths absorbed more strongly and having a higher quantum yield. The consequence, as will be seen, is that the $\varepsilon(\lambda)$ for chemical processes declines sharply with wavelength.

A complication in analyzing UV photochemistry is that the details of what goes on between photon absorption and the resultant product, *e.g.* carbon dioxide, is poorly understood. Marine CDOM is a complex mixture of compounds and it is not clear which components actually participate in photochemistry. Measurements can only determine 'apparent' quantum yields. This complication is further compounded in organisms, where the targets of UV are imbedded in organized structures that can also absorb and scatter UV before it reaches a sensitive molecule. Despite our imprecise knowledge of the mechanisms involved in UV effects, basic photochemical principles are assumed to apply within biological tissues. The most important biomolecules (proteins and nucleic acids) exhibit increasing absorbance with decreasing wavelength.[13] Other molecules vary in their absorption spectrum, particularly pigments and light receptors, yet the overall cellular absorption spectrum generally declines with increasing wavelength through the UV.[13] Thus the theoretical expectation, and empirical observation, is that biological effects of UV also have strong wavelength dependence.

Temporal Dependence

Solar ultraviolet radiation varies with time as well as wavelength, so quantitative models of UV effects need to take into account temporal as well as spectral dependence. The simplest case is a first-order photochemical reaction for which the photoproduct will increase with increasing exposure. If there is little depletion of the reactant pool, then product formation will be proportional to the total number of photons absorbed, independent of intensity or duration. The term reciprocity is applied when effects are dependent only on cumulative exposure. In reality, reactants eventually become limiting and products may be unstable. If competing reactions are significant, accumulation of a product may be limited to a steady-state concentration, which will be a function of irradiance.

As for photochemical reactions, the temporal characteristics of biological responses to UV also vary, depending on the relative strength of damage and repair processes. The rate of damage is analogous to the forward reaction rate, whereas 'repair', *i.e.* restoration or reactivation of function, is analogous to a back-reaction. Reciprocity will apply if repair is not significant over the time scale of exposure. However, in many cases, repair processes are important and effects over a sufficiently long period will be dependent on irradiance.[8]

The preceding discussion suggests that many general principles and characteristics apply to both chemical and biological effects of UV, and that it could be

---

[12] R. Whitehead and S. J. de Mora, this volume, Chapter 3.
[13] F. Garcia-Pichel, *Limnol. Oceanogr.*, 1994, **39**, 1704.

instructive to compare the approaches and results obtained in each field of study. However, there are many unique aspects of each field, and these aspects will be considered in the next two sections before returning to a more comparative discussion.

2 Chemical Action Spectra

The action spectrum is a fundamental parameter used to characterize the spectral dependence of a photochemical or photophysical process. For seawater, or other natural waters which are primarily composed of poorly defined chromophores, the action spectrum, $\varepsilon(\lambda)$, is the wavelength-dependent product of the apparent quantum yield, $\Phi(\lambda)$, and the absorbance coefficient, $a(\lambda)$. Although there are numerous published wavelength-dependent quantum yields in seawater,[12,14,15] the corresponding action spectra are presented in comparatively few cases.[16] Action spectra for carbonyl sulfide, hydrogen peroxide, and carbon monoxide photoproduction all show an exponential decrease with increasing wavelength from 290 to 400 nm, with considerably less variability than observed among different biological action spectra.[17] Ideally, if the action spectrum for a photoprocess is known and does not exhibit significant spatiotemporal variations in seawater, then it will be possible to predict the rate for this process with accurate knowledge of the spectral irradiance. The issue of reciprocity should be tested, however, especially for long-term, multi-day photon exposures, owing to uncertainties regarding reaction kinetics (*e.g.*, loss of precursors). Despite evidence suggesting that one or a few action spectra may be enough to describe a marine photoprocess,[16] there is no compelling evidence to conclude *a priori* that action spectra for different oceanic environments should be the same. Given the regional differences in the sources (*e.g.*, riverine, autochotonous) and cycling (*e.g.*, photobleaching, microbial degradation) of chromophoric dissolved organic matter (CDOM) that are observed in seawater, it is surprising that action spectra are the same in contrasting oceanic environments (*e.g.*, coastal *vs.* oligotrophic). Observed similarities may be fortuitous, and in some cases based on too few measurements. For hydrogen peroxide photoproduction, the slope of a log-linear fit to the action spectra in environments ranging from oligotrophic to high CDOM coastal waters varied from $0.04\,\text{nm}^{-1}$ to $0.07\,\text{nm}^{-1}$ at 25°C, with an average slope of $0.05\,\text{nm}^{-1}$ (SD $\pm\ 0.01\,\text{nm}^{-1}$).[18] Similarly, photobleaching (loss of absorbance), which is at least partially the consequence of UV degradation of dissolved organic matter, might be expected to have a comparable spectral dependence. However, action spectra for photobleaching of lake water have a much shallower slope, on the order of $0.017\,\text{nm}^{-1}$.[19,20] Action spectra for the

[14] W. L. Miller and S. C. Johannessen, presented at the 2000 Ocean Science Meeting, San Antonio, TX, 2000, OS93.
[15] B. H. Yocis, D. J. Kieber, and K. Mopper, *Deep-Sea Res.*, 2000, **47**, 1077.
[16] M. A. Moran and R. G. Zepp, *Limnol. Oceanogr.*, 1997, **42**, 1307.
[17] M. A. Moran and R. G. Zepp, in *Microbial Ecology of the Oceans*, ed. D. L. Kirchman, Wiley, New York, 2000, p. 201.
[18] G. W. Miller, *Wavelength and Temperature Dependent Apparent Quantum Yields for Photochemical Formation of Hydrogen Peroxide In Seawater*, Masters Thesis, State University of New York, College of Environmental Science and Forestry, Syracuse, 2000.

photochemical formation of dissolved inorganic carbon also showed an exponential decrease with increasing wavelength.[21] Inshore waters were higher, on average, by nearly a factor of two relative to coastal seawater and open ocean samples at 290 nm. In contrast, coastal and open ocean waters were nearly the same at this wavelength. Action spectra for these three different water types converged with increasing wavelength until they were nearly indistinguishable at wavelengths longer than approximately 350 nm.

Errors in experimentally determined quantum yields and absorption coefficients ultimately determine uncertainties in action spectra. Both sources of error will be spectrally dependent and increasing at longer wavelengths, approaching their respective detection limits between 380 and 450 nm in oligotrophic, low CDOM waters. For example, at the Hawaii station ALOHA, an optically clear open ocean site, the percent coefficient of variation for the determination of $a(\lambda)$ was 8% at 320 nm and 66% at 400 nm, based on spectral scans of five separate seawater aliquots. For the same sample, the coefficient of variation in the determination of $\Phi(\lambda)$ was 4% and 91% at 320 and 400 nm, respectively, determined from propagation of errors analysis of equation (2) (*vide infra*). For action spectra that extend out beyond 400 nm, individual errors for $\Phi(\lambda)$ and $a(\lambda)$ are expected to be even larger, especially in optically clear waters that comprise most of the world's oceans. In the extreme case, it may not be possible to quantify either $\Phi(\lambda)$ or $a(\lambda)$. This was the case for the photolysis of dimethyl sulfide (DMS). It was not possible to determine wavelength-dependent quantum yields or absorption coefficients at wavelengths greater than 400 nm in equatorial Pacific seawater even though DMS photolysis occurred out to 580 nm.[22] Despite this limitation, it is still possible to define an action spectrum for DMS photolysis that is specific to this seawater. In optically clear waters such as the equatorial Pacific, the blue/visible spectral region is well below the detection limit for determination of $a(\lambda)$ by conventional spectroscopy.[23] Experimental errors in photochemical action spectra will be compounded when extrapolated to the determination of environmental rates (equation 1) in the water column because rates will be red-shifted with depth owing to the faster attenuation of shorter wavelengths (see ref. 16, p. 1312). The increasing importance of the longer wavelengths will result in larger uncertainties deeper in the water column. A further complication is that absorbance coefficients vary as a function of depth,[24] which is problematic when applying surface action spectra to *in situ* rates. However, there are very few published *in situ* spectral absorbance measurements to gauge the potential magnitude of this error. Thus, one of the major limitations in chemical action spectra that will challenge oceanographers is the precise determination of absorbance coefficients in the visible part of the spectrum. In the laboratory,

[19] C. L. Osburn, H. E. Zagarese, W. Cravero, D. P. Morris, and B. R. Hargreaves, *Limnol. Oceanogr.*, 2000, submitted.
[20] C. L. Osburn, *Photochemical Changes in the Dissolved Organic Carbon of Lakes: Implications for Organic Carbon Cycling*, Masters Thesis, Lehigh University, Bethlehem, PA, 2000.
[21] S. C. Johannessen and W. L. Miller, personal communication, 2000.
[22] D. J. Kieber, R. P. Kiene, and T. S. Bates, *J. Geophys. Res.*, 1996, **101**, 3717.
[23] S. A. Green and N. V. Blough, *Limnol. Oceanogr.*, 1994, **39**, 1903.
[24] S. R. Beaupre, D. J. Kieber, and K. Mopper, presented at the 2000 Ocean Sciences Meeting, San Antonio, TX, 2000, OS128.

significant advances in the measurement of extremely low absorbance coefficients, well below $0.1\,\text{m}^{-1}$, may be realized in the near future owing to the development of capillary waveguide (CW) technology, which has been recently applied to seawater absorbance measurements.[25] The advantage of the CW is that long pathlengths can be used (ca. 0.5 m). Therefore, lower detection limits are theoretically possible. The disadvantage, though, is that baseline offsets relative to the blank are more pronounced than observed using conventional spectroscopic techniques; until this problem is resolved, spectroscopic measurement of optically clear waters by CW technology will be no more sensitive than conventional techniques.

In addition to experimental uncertainties for $\Phi(\lambda)$ and $a(\lambda)$, there are other errors associated with using action spectra to determine photochemical rates from equation (1). Implicit assumptions in rate calculations are that $\Phi(\lambda)$ and $a(\lambda)$ do not change with cumulative exposure. Clearly this will not be the case for $a(\lambda)$ because photobleaching will occur [i.e., $a(\lambda)$ will decrease], and photobleaching will be spectrally dependent.[17,26] The degree of photobleaching, and hence the extent of this error, will be a function of exposure. As an example, solar exposure of surface oligotrophic seawater during the summer in the northwest Atlantic Ocean will result in approximately 10–30% bleaching. The extent of bleaching will decrease with depth in the water column down to approximately 4 m.[24] Many other investigators have observed this rate of bleaching in surface waters in both marine and freshwater environments.[27,28] Variations in $a(\lambda)$ with depth and cumulative exposure are the main reasons why photochemists report wavelength-dependent quantum yields in the literature and not action spectra.

The effect of cumulative exposure on photochemical quantum yields is not as straightforward as for $a(\lambda)$. For primary photochemical processes where the chromophores involved are known, quantum yields should be constant in seawater at low precursor concentrations as long as there is no appreciable buildup of products or transient species (e.g., superoxide anion) that change reaction conditions. However, this will not necessarily be the case when determining apparent quantum yields in seawater:

$$\Phi(\lambda) = \frac{R(\lambda)}{E(\lambda)(1 - 10^{-A(\lambda)})} \qquad (2)$$

where $\Phi(\lambda)$ is the wavelength-dependent apparent quantum yield (mol mol photons^{-1}), $R(\lambda)$ is the rate of production or loss of a chemical species (mol s^{-1}), $E(\lambda)$ is the spectral radiant flux determined from chemical actinometry (mol photons^{-1}) [note that $E(\lambda)$ is equivalent to $P(\lambda)$ defined by the photochemistry IUPAC committee[29]], $A(\lambda)$ is the absorbance, and $(1 - 10^{-A(\lambda)})$ is the fraction of

---

[25] E. J. D'Sa, R. G. Steward, A. Vodacek, N. V. Blough, and D. Phinney, *Limnol. Oceanogr.*, 1999, **44**, 1142.
[26] K. Mopper and D. J. Kieber, unpublished data, 2000.
[27] K. Mopper and D. J. Kieber, in *The Effects of UV Radiation on Marine Ecosystems*, ed. S. de Mora, S. Demers, and M. Vernet, Cambridge University Press, New York, 2000, p. 101.
[28] P. S. Kuhn and W. L. Miller, presented at the 2000 Ocean Sciences Meeting, San Antonio, TX, 2000, OS112.
[29] J. W. Verhoeven, *Pure Appl. Chem.*, 1996, **68**, 2223.

monochromatic radiation absorbed by 0.2 μm-filtered seawater. In this equation, if $R(\lambda)$ decreases faster than photobleaching [*i.e.*, faster than $(1 - 10^{-A(\lambda)})$ decreases], then quantum yields will decrease over time. Conversely, if $R(\lambda)$ decreases more slowly than photobleaching, then apparent quantum yields will increase. Only when $R(\lambda)$ and the fraction of absorbed radiation vary at the same rate with cumulative exposure, will apparent quantum yields remain constant. This scenario is unlikely in seawater for most photoprocesses once photobleaching has occurred because the precursors for each photoprocess presumably are only a fraction of CDOM and there is no evidence that would suggest that precursors and CDOM should decrease at the same rate. Apparent quantum yields for the photochemical loss of dissolved oxygen were extremely dose dependent in seawater, roughly decreasing threefold at 310 nm then leveling off to a constant value at longer irradiations.[30] Similarly, H_2O_2 quantum yields decreased with increasing absorbed dose.[30] Thus, in order to determine environmentally relevant rates for photochemical processes, equation (1) will have to incorporate additional terms to account for temporal as well as spectral changes in $\Phi(\lambda)$ and $a(\lambda)$ as they affect $\varepsilon(\lambda)$. Ultimately, photochemistry will be a nonlinear function of UV exposure under some conditions. A similar situation occurs in many cases for biological effects, as will shown in the next section.

3 Biological Weighting Functions

General Characteristics

Like photochemical action spectra, the spectral dependence of the biological effects of UV shows an exponential-like decline through the UV-B and UV-A.[8,9] Unlike photochemistry, it is difficult to partition this steep spectral dropoff between a gradient in absorbance and spectral variation in quantum yield. Cells absorb and scatter light, with significant effects on transmission of radiation to intracellular chromophores, even for marine microorganisms.[13,31] For example, the action spectrum is known for UV damage of DNA by formation of thymine dimers,[32] yet it is difficult to determine cellular DNA damage *a priori* because of uncertainty in how much UV radiation reaches the nucleus. The reactivity and absorbance of the chromophore(s) may also be affected by being bound to a large complex *in vivo*.[33] Both DNA and proteins have declining absorbance with increasing wavelength, but the actual chromophore(s) (and associated absorbance spectra) for many types of UV damage are either uncertain or unknown. Thus, there are few cases where spectral shape *per se* is diagnostic of a specific chromophore.[8]

UV effects on living systems are generally expressed in terms of a simple weight, $\varepsilon(\lambda)$, and not the product of an (apparent) quantum yield and absorbance. Effects are therefore expressed as a function of incident, as opposed to absorbed,

---

[30] S. S. Andrews, S. Caron, and O. C. Zafiriou, *Limnol. Oceanogr.*, 2000, **45**, 257.
[31] A. Morel, in *Particle Analysis in Oceanography*, ed. S. Demers, Springer, Berlin, 1990, p. 141.
[32] R. B. Setlow, *Proc. Natl. Acad. Sci. USA*, 1974, **71**, 3363.
[33] W. F. Vincent and P. J. Neale, in *The Effects of UV Radiation on Marine Ecosystems*, ed. S. J. de Mora, S. Demers, and M. Vernet, Cambridge University Press, Cambridge, 2000, p. 149.

radiation. Responses between organisms (*e.g.* species of phytoplankton) vary, and part of this variation can arise simply from differences in absorbance. However, the variability in the absorbance of cellular matter is quite modest,[13] at least in comparison to the variability in CDOM absorbance. An important experimental consideration in determination of $\varepsilon(\lambda)$ is that exposures should be as optically thin as possible so that incident irradiance is accurately known.

Dose–Response for Biological Effects: Importance of Repair

Organisms have developed many mechanisms to minimize and counteract the effects of UV exposure. The reader is referred to some recent reviews for the biological details;[33] here we explore how the balance between damage and repair influences modeling of UV effects. The term 'repair' is used in a generic sense as all those processes that restore or reactivate function. This includes anything from simple photoenzymatic reversal of photochemical reactions, such as splitting of thymine dimers, to a complex cycle consisting of degradation of damaged complexes, synthesis of precursors and reassembly of new, replacement complexes. The presence of these processes affects the time course of effects (*e.g.* magnitude of inhibition) during exposure and thus how weighted exposure (equation 1) is translated into overall effect. A function describing the variation of effect with exposure is called an 'exposure response curve' (ERC). A concise introduction to ERCs is given below; more detail on their derivation and application is provided in previous reviews.[8,9,34] In the absence of repair, the time course of effects follows an exponential 'survival curve':[35]

$$P(t) = P_0 e^{-H^*}$$

$$H^* = \int_0^\tau E^* \, dt$$

(3)

where $P(t)$ is some measure of organismal activity (*e.g.* photosynthesis), initially at level P_0, after a period, τ, during which the integrated weighted exposure is H^*. On the other hand, if repair is active and significant, the loss of activity will eventually arrive at a steady state that is a function of the ratio of damage (k) and repair (r) rates (s^{-1}):

$$\frac{P}{P_0} = \frac{r}{(r+k)} = \frac{1}{\left(1 + \dfrac{k}{r}\right)}$$

(4)

The hyperbola specified by equation (4) described the steady-state rates of algal photosynthesis in a laboratory culture under UV and photosynthetically available radiation (PAR 400–700 nm) exposure, consistent with a dynamic balance between damage and repair.[36] Equation 4 applies when the repair rate increases as damage accumulates. A rectangular hyperbola model applies when repair is constant, independent of cumulative damage.[8] In either case, the

---

[34] T. P. Coohill, in *Stratospheric Ozone Depletion/UV-B Radiation in the Biosphere*, ed. R. H. Biggs and M. E. B. Joyner, Springer, Berlin, 1994, p. 57.
[35] W. Harm, *Biological Effects of Ultraviolet Radiation*, Cambridge University Press, Cambridge, 1980.
[36] M. P. Lesser, J. J. Cullen, and P. J. Neale, *J. Phycol*, 1994, **30**, 183.

transition to steady state generally follows first-order kinetics, *e.g.*

$$\frac{P(t)}{P_0} = \frac{r}{(k+r)} + \frac{k}{(k+r)} e^{-(k+r)t} \quad (5)$$

Which ERC is used to estimate the effect of exposure for a specific process depends on the time scale of the assessment relative to the inherent dynamics of the system. If repair is active and the exposures are sufficiently long to attain steady state, the correct approach is to weight irradiance [$E(\lambda)$, mW m^{-2} nm^{-1}] and predict response as a hyperbolic function of weighted irradiance. This approach was taken to develop an integrated model of photosynthetic response to UV and PAR, *i.e.* the biological weighting function, photosynthesis *vs.* irradiance (BWF-PI) model, with weightings, $\varepsilon(\lambda)$, having units of reciprocal mW m^{-2} so that weighted exposure for inhibition E^*_{inh} (equation 1) is dimensionless.[37] The BWF-PI model was modified for the case when repair is not active (equation 3 applies) and the objective was to predict the average photosynthetic rate over a specific exposure period.[38] In this case, weighted exposure for inhibition, H^*_{inh}, is obtained by weighting radiant exposure [$H(\lambda)$, J m^{-2} nm^{-1}] similarly as for E^*_{inh}:

$$H^*_{inh} = \sum_{\lambda=280nm}^{400nm} \varepsilon_H(\lambda) \cdot H(\lambda) \cdot \Delta\lambda \quad (6)$$

where $H(\lambda) = \int E(\lambda) \cdot dt$. The weightings, ε_H (reciprocal J m^{-2}), define the effects of cumulative exposure in contrast to irradiance. This second model is termed the BWF$_H$-PI model because of the dependence on H, compared to the earlier BWF$_E$-PI model (with weights ε_E). Intermediate approaches can be formulated in those cases where repair is still significant but steady state is not reached within the time period of interest, in which case response is dependent on both cumulative exposure and intensity. For example, a time-dependent approach can be implemented using equation (5) in which damage is defined by a BWF$_H$ and repair is represented by a separate parameter (r). These examples illustrate the interaction between spectral and temporal responses to UV exposure (see also Neale[8]). To accurately define models of biological responses to UV exposure, observations will usually be needed of the variation in UV effects in relation to both varying spectral composition (for the BWF) and exposure times (for the ERC).

Measurement Approaches

The activity of many repair mechanisms in microorganisms, such as photosynthesis and photoenzymatic repair, is dependent upon concurrent exposure to UV-A and PAR. Thus, measurements of UV response to monochromatic exposures may not result in environmentally relevant biological weighting functions. To be more representative of biological responses to environmental UV, measurements are made using polychromatic exposures composed of a range of wavelengths. The basic approach is to generate a set of spectra using cutoff filters, *i.e.* filters

---

[37] J. J. Cullen, P. J. Neale, and M. P. Lesser, *Science*, 1992, **258**, 646.
[38] P. J. Neale, J. J. Cullen, and R. F. Davis, *Limnol. Oceanogr.*, 1998, **43**, 433.

which pass longer wavelength light starting at successively shorter cutoff wavelengths.[11] The tradeoff for greater realism is that effects can no longer be precisely attributed to specific wavelengths. Instead, the weights are composites of the effects at that wavelength and the interactive effects of other wavelengths.[39]

Experimental UV exposures may be accomplished using filtered incident solar irradiance alone, solar irradiance supplemented by UV lamps, or solar simulator (xenon arc) lamps. Cutoff filters (Schott glass, various types of other glasses and plastics) control the spectral composition and neutral density screens control the overall intensity of the treatments. Solar simulators can approximate the spectral distribution of solar irradiance, but there is no lamp/filter combination that exactly reproduces solar irradiance. Spectral features of solar irradiance that are difficult to simulate are (1) the sharp drop in energy with wavelength in the UV-B (Chapter 2, this volume), and (2) the high ratio of UV-A and PAR to UV-B. Furthermore, no artificial source can simulate the changes of spectral irradiance with depth in surface waters with various constituents.[40] Fluorescent UV lamps, which have wide commercial availability due to their use as tanning lamps, have much higher amounts of short wavelength *vs.* long wavelength UV-B than found in solar irradiance, even after filtering through cellulose acetate sheets to remove the UV-C component.[9] Since the BWF weight at any wavelength implicitly includes interactions with other wavelengths, there can be some question about the capability of a BWF that was estimated using artificial sources to predict responses to solar irradiance. On the other hand, use of solar irradiance alone provides little reliable variation in the shortest UV-B wavelengths (<305 nm), though the rotation of the 'ozone hole' provides a natural source of spectral variation in Antarctica.[41] Variability in cloud cover also complicates experiments with natural irradiance. Questions about the predictive power of a BWF measured with any source can be addressed by comparing BWF-based predictions to observations of UV effects using an independent set of response–irradiance measurements.[42] Further discussion of methodological considerations is given by Neale[8] and Cullen and Neale.[9]

Determination of BWFs for inhibition of phytoplankton photosynthesis by UV has used a special incubator known as the 'photoinhibitron'. Measurements of photosynthesis (incorporation of ^{14}C-labeled dissolved inorganic carbon) are made under 72–80 polychromatic UV-PAR treatments during 1–2 h exposures. Two designs have been developed, one with 72 1-cm diameter cuvettes (1–2 mL volume),[37,43] and a second with 2.5-cm diameter quartz cuvettes (5–10 mL), all simultaneously illuminated by filtered output from a Xe lamp. The second setup is optimized for Antarctic phytoplankton with larger cells ($>100\,\mu$m). The large

[39] T. P. Coohill, *Photochem. Photobiol.*, 1991, **54**, 859.
[40] J. T. O. Kirk, B. R. Hargreaves, D. P. Morris, R. B. Coffin, B. David, D. Frederickson, D. Karentz, D. R. S. Lean, M. P. Lesser, S. Madronich, J. H. Morrow, N. B. Nelson, and N. M. Scully, *Arch. Hydrobiol. Beih Ergebn. Limnol.*, 1994, **43**, 71.
[41] R. C. Smith, B. B. Prézelin, K. S. Baker, R. R. Bidigare, N. P. Boucher, T. Coley, D. Karentz, S. MacIntyre, H. A. Matlick, D. Menzies, M. Ondrusek, Z. Wan, and K. J. Waters, *Science*, 1992, **255**, 952.
[42] M. P. Lesser, P. J. Neale, and J. J. Cullen, *Mol. Mar. Biol. Biotechnol.*, 1996, **5**, 314.
[43] P. J. Neale, M. P. Lesser, and J. J. Cullen, in *Ultraviolet Radiation in Antarctica: Measurements and Biological Effects*, ed. C. S. Weiler and P. A. Penhale, American Geophysical Union, Washington, 1994, p. 125.

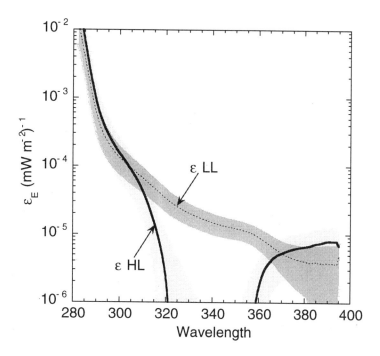

Figure 1 Example BWFs for high light (HL, solid line) and low light (LL, broken line) grown *Gymnodinium sanguineum* (same data set as discussed in Neale et al.[49]). Each BWF was estimated from data from a single photoinhibitron experiment based on an ERC of $1/(1 + E^*_{inh})$. Shaded area shows the 95% confidence interval for each BWF

volume setup has also been used for polychromatic exposures of filtered water for determination of photochemical production. Simultaneous exposure of many spectral treatments makes possible the definition of spectral response much more rapidly than the usual approach of serial exposures to monochromatic irradiance. Polychromatic exposures may also reveal interactions between photochemical reactions induced by UV at different wavelengths; however, initial results indicate a close agreement between monochromatic and polychromatic based estimates of a hydrogen peroxide action spectrum.[18]

Using photoinhibitrons, BWFs have been defined for cultures and natural populations in a range of environments (see Section 4). High-resolution spectral irradiance is measured for each treatment, and the resultant set of spectra are analyzed using principal component analysis (PCA).[9,37,43] Scores representing the spectral variation in each treatment are used in a nonlinear regression to estimate the BWF-PI parameters. This procedure generates a BWF that usually accounts for more than 90% of the variation in the data. The GSL confidence intervals in the estimated $\varepsilon(\lambda)$ are on order of 30–50%, assuming that the proper number of components have been identified (Figure 1).

A similar polychromatic apparatus with 34 positions and seven cutoffs has been used to estimate BWFs for UV effects on the survival and development of fish and zooplankton eggs.[44] Polychromatic exposures have also been conducted for phytoplankton photosynthesis under natural solar illumination.[45] In these

[44] J. H. M. Kouwenberg, H. I. Browman, J. J. Cullen, R. F. Davis, J.-F. St-Pierre, and J. A. Runge, *Mar. Biol.*, 1999, **134**, 269.
[45] N. P. Boucher and B. B. Prézelin, *Mar. Ecol. Prog. Ser.*, 1996, **144**, 223.

experiments, the BWFs were derived by assuming that $\varepsilon(\lambda)$ follow a generalized nth order exponential.[9,46] Usually a first- or, at most, second-order exponential function is sufficient.

4 Comparative Spectroscopy of Weighting Functions

With the tools for quantitative assessment now in hand, there are an increasing number of chemical action spectra and BWFs being defined for aquatic processes. Weighting functions for inhibition of photosynthesis by UV have been emphasized, given the central role of this process in the global carbon cycle. Lacking other alternatives, initial assessments of the effect of O_3 depletion on photosynthesis used a weighting function describing inhibition of electron transport in chloroplast membranes, with recognized limitations.[47,48] However, the initial application of the BWF_E-PI model to cultures of diatoms and dinoflagellates showed that phytoplankton photosynthesis was less sensitive to UV-A than inhibition of chloroplast electron transport.[37] On the other hand, inhibition of photosynthesis was not exclusively a UV-B effect, as for photodamage of DNA. In a follow-up study with high and low light grown dinoflagellates, careful comparison between BWFs revealed the optical effect of cellular accumulation of UV absorbing compounds (mycosporine amino acids, MAAs) (Figure 1).[49] More comparative studies are underway using cyanobacteria, chlorophyte, and cryptophyte cultures grown under a variety of experimental conditions.

The culture studies have led, and continue to lead, to important insights on how UV sensitivity varies both as a function of taxonomic group and growth conditions. However, an even more important use of the photoinhibitron has been in measuring responses to UV by natural phytoplankton, particularly Antarctic assemblages (Figure 2). The initial deployment was made in waters adjacent to McMurdo Station (78°S 166°E), at the southern margin of the Ross Sea. Phytoplankton growth in nearby, but inaccessible, open water was simulated in tanks exposed to natural solar irradiance. Photosynthesis was saturated under growth irradiance, so these cultures were acclimated to higher light and were fairly resistant to UV. The growth conditions were artificial, however, and subsequent studies focused on sampling natural assemblages in critical habitats directly exposed to the Antarctic 'ozone hole'. These areas include the Weddell–Scotia Confluence (60°S 51°W), an area of high biomass during spring-time ozone depletion, and the near-shore areas of the Antarctic Peninsula around Palmer Station (64°S 64°W). In most cases, these natural assemblages were much more sensitive to UV than the McMurdo cultures (Figure 2). For samples at Palmer Station in 1997, the BWFs partitioned into two distinct groups. The high sensitivity group was found mainly during periods with partial pack-ice cover and responses to 1 h exposures were comparable to that of the Weddell–Scotia phytoplankton.[50,51] The low sensitivity group occurred

---

[46] R. D. Rundel, *Physiol. Plant.*, 1983, **58**, 360.
[47] R. C. Smith, K. S. Baker, O. Holm-Hansen, and R. S. Olson, *Photochem. Photobiol.*, 1980, **31**, 585.
[48] R. C. Smith and K. S. Baker, *Photochem. Photobiol.*, 1979, **29**, 311.
[49] P. J. Neale, A. T. Banaszak, and C. R. Jarriel, *J. Phycol.*, 1998, **34**, 928.

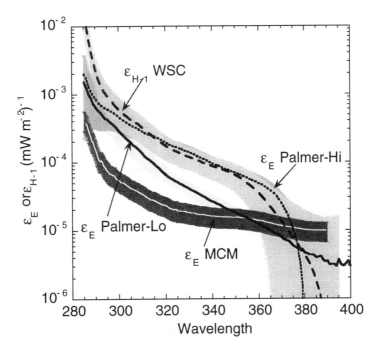

Figure 2 Average BWFs for natural phytoplankton assemblages in different Antarctic marine environments. The McMurdo (MCM) BWF (inverse mW m^{-2}) was estimated using the responses from two photoinhibitron exposures using the ERC, $1/(1 + E^*_{inh})$. The Weddell–Scotia Confluence (WSC) BWFs are the average of six BWFs determined in the austral spring (Oct–Nov) of 1993, using the ERC $(1 - \exp[-H^*_{inh}])/H^*_{inh}$,[38] and converted to units of inverse mW-h m^{-2} by multiplying by 2.73.[51] The Palmer Station BWFs[50,51] were fit using the ERC $1/(1 + E^*_{inh})$. Shaded bands indicate standard deviations except for the McMurdo BWF in which they indicate an estimated 95% confidence interval. Where bands overlap, a line indicates the bound of the 'hidden' band

during episodes of open water and approached the lower weights of the BWF estimated for the McMurdo cultures. Preliminary results from additional field work conducted during 1998 and 1999 confirmed these trends.[50]

The Antarctic studies and concurrent studies in a temperate estuary (Rhode River[52]) have revealed variability in sensitivity to UV and diversity in ERCs of natural phytoplankton assemblages. Photosynthesis by McMurdo, Palmer, and Rhode River assemblages attained a steady-state level during UV and PAR exposure, and thus photosynthesis was modeled using the BWF$_E$-PI approach. In sharp contrast, recovery processes made a negligible contribution to the response of phytoplankton sampled from the deeply mixed waters of the Weddell–Scotia Confluence (WSC). Consequently, the BWF$_H$-PI model was applied in which the cumulative effect of inhibition is described as the integral of a semi-logarithmic survival curve. The lack of repair may be a byproduct of the generally lower metabolic activity of phytoplankton acclimated to the low light environment of the WSC (because of deep mixing).[38] Further discussion of these models and their application can be found.[8,9] Studies in progress will examine the responses of phytoplankton in other marine and freshwater environments.

The variation in ERCs used to describe the effect of UV on photosynthesis poses a challenge for comparing BWFs of different phytoplankton assemblages. The $\varepsilon(\lambda)$ are fit with different units and thus cannot be directly compared. The problem can be avoided by normalizing the $\varepsilon(\lambda)$ to weight at some wavelength (e.g. 300 nm), but then comparisons of sensitivity are not possible. However,

---

50 J. J. Fritz, P. J. Neale, R. F. Davis, and A. T. Banaszak, presented at the 2000 Ocean Sciences Meeting, San Antonio, TX, 2000, OS199.
51 P. J. Neale, J. J. Fritz, and R. F. Davis, *Rev. Chil. Historia Natural*, 2000, in press.

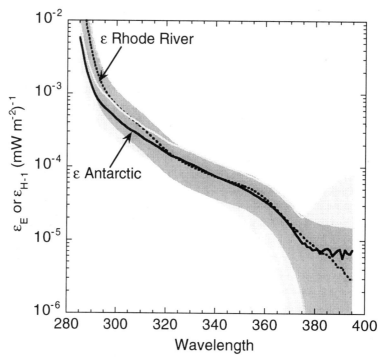

Figure 3 Comparison of BWFs for natural phytoplankton assemblages from Antarctica, and a temperate estuary (Rhode River, Chesapeake Bay, USA). The Antarctic BWF is the mean ($\pm$ standard deviation) of all BWFs shown in Figure 2. The Rhode River BWF is the mean ($\pm$ standard deviation) of BWFs estimated from photoinhibitron measurements made at approximately monthly intervals during 1995–1996.[52] Shaded bands indicate standard deviations. Where bands overlap, a line indicates the bound of the 'hidden' band

absolute comparisons can still be made at a specific time scale. Conversion factors have been derived for instantaneous (time 'zero') and 1-h exposures.[51] The comparisons over a 1-h time scale are shown here. This is the most conservative comparison since 1 h is the duration of experimental exposure. The BWF_H are multiplied by 2.73, which adjusts for the relative differences between ERCs at a given exposure (see caption, Figure 2) and the conversion between joules and milliwatt-hours.[51]

Within environments, BWFs vary by 2- to 3-fold at any given wavelength between samples separated in space or time (Figures 2, 3). However, the average BWF over many samples appears similar between environments (Figure 3). Phytoplankton in a temperate estuary had the same average sensitivity to UV as the average of the Antarctic assemblages (when scaled for effects over 1 h exposure).[52] It is unknown whether this pattern would also be seen in phytoplankton from the surface layer of warm (temperate vs. tropical) oceans. In each of these environments, assemblages are occasionally encountered which have low sensitivity to UV, though it is usually difficult to partition the source of such variation between changes in species composition and physiological adaptation. On average, though, natural populations are much less resistant to UV than are cultures grown under nutrient sufficient, benign light conditions. The average BWF for all marine natural assemblages (Antarctic and temperate) is consistently greater than average weights for cultures of diatoms and dinoflagellates, especially in the UV-A (Figure 4). Preliminary BWFs for other species of diatoms, dinoflagellates, and cryptophytes[53] are generally consistent with these earlier

[52] A. T. Banaszak and P. J. Neale, *Limnol. Oceanogr.*, 2000, in press.
[53] E. Litchman, personal communication, 2000.

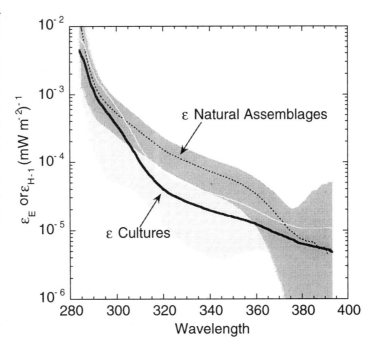

Figure 4 Comparison of BWFs for natural assemblages and nutrient sufficient phytoplankton cultures. The natural assemblage BWFs are the mean of the BWFs in Figure 3. The culture BWFs are the mean (± standard deviation) for *Phaeodactylum*, *Prorocentrum*[37] and *Gymnodinium*.[49] Shaded bands indicate standard deviations. Where bands overlap, a line indicates the bound of the 'hidden' band

studies. This suggests that some factor or combination of factors (nutrient, light limitation, *etc.*) constrains the full expression of UV defense mechanisms in natural assemblages. For example, growth under nitrogen limitation increases the sensitivity of dinoflagellate cultures to inhibition by UV, shifting the BWF closer to the natural assemblage average.[54]

While much effort has been devoted to analyzing effects on marine photosynthesis, it is unclear how much these results can be generalized to other marine processes. Marine bacteria receive more UV exposure to cellular matter than phytoplankton because of their smaller cell size.[13] Bacterioplankton incubated under near-surface solar intensities show UV specific decreases in assimilation and increases in DNA damage (thymidine dimers).[55] Solar exposure of bacterioplankton accounts for most of the near-surface accumulation of thymidine dimers in planktonic environments.[56] These observations have motivated efforts to define BWFs for the response of marine bacteria to UV using the photoinhibitron approach.[57] Preliminary results show differences between effects of UV on nucleic acids synthesis (thymidine assimilation) *versus* protein synthesis (leucine assimilation), with the former being more sensitive to UV-A. A similar pattern is observed in incubations under solar illumination.

[54] E. Litchman and P. J. Neale, presented at the American Society of Limnology and Oceanography Aquatic Sciences Meeting, Santa Fe, NM, 1999, p. 110.
[55] W. H. Jeffrey, P. Aas, M. M. Lyons, R. Pledger, D. L. Mitchell, and R. B. Coffin, *Photochem. Photobiol.*, 1996, **64**, 419.
[56] W. H. Jeffrey, R. J. Pledger, P. Aas, S. Hager, R. B. Coffin, R. Von Haven, and D. L. Mitchell, *Mar. Ecol. Prog. Ser.*, 1996, **137**, 283.
[57] J. A. Peloquin, P. J. Neale, W. H. Jeffrey, and J. Kase, presented at the 2000 Ocean Sciences Meeting, San Antonio, TX, 2000, paper OS11.

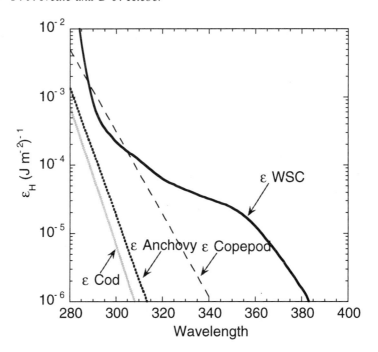

Figure 5 BWFs for UV effects on fish and zooplankton egg survival compared to the average BWF for inhibition of photosynthesis in the Weddell–Scotia Confluence (WSC).[38] All these BWFs assume that repair is not significant so that response follows an exponential 'survival curve' and $\varepsilon(\lambda)$ has units of inverse $J\ m^{-2}$. The cod[44] and copepod[61] BWFs were fit by a single exponential slope. The anchovy spectrum is the modified Setlow DNA spectrum scaled for the reported exposure for 50% mortality in a 12-day incubation[58]

Some species of zooplankton (small crustaceans and the larval stages of many marine organisms) spend portions of their life cycle in near-surface waters and thus may receive significant UV exposure.[58–60] The effects are usually observed as increases in probability of mortality and developmental abnormalities. The few BWFs for these effects show a strong dependence on UV-B similar to the action spectrum for damage to DNA (Figure 5). At high exposures, effects are best predicted from cumulative exposure. Cod eggs (*Gadus morhua*)[44] and anchovy larvae (*Engraulis mordax*)[58] were much less sensitive to UV than eggs of a copepod, *Calanus finmarchicus*.[61] The copepod was sensitive to even short exposures to surface UV, and decreased survival due to UV-B exposure scaled similarly as inhibition of photosynthesis in the WSC phytoplankton which lacked repair. There has been limited work on determining BWFs for UV effects on growth rate in phytoplankton, but these results also suggest a similarity of response to the action spectrum for DNA damage and a large variation in sensitivity between species.[62,63]

Comparison between Chemical Action Spectra and BWFs

Chemical and biological effects of UV have a common base in the photochemical

[58] J. R. Hunter, S. E. Kaupp, and J. H. Taylor, *Photochem. Photobiol.*, 1981, **34**, 477.
[59] J. R. Hunter, J. H. Taylor, and H. G. Moser, *Photochem. Photobiol.*, 1979, **29**, 325.
[60] D. M. Damkaer, in *The Role of Solar Ultraviolet Radiation in Marine Ecosystems*, ed. J. Calkins, Plenum Press, New York, 1982, p. 701.
[61] J. H. M. Kouwenberg, H. I. Browman, J. J. Cullen, R. F. Davis, J.-F. St-Pierre, and J. A. Runge, *Mar. Biol.*, 1999, **134**, 285.
[62] A. G. J. Buma, A. H. Engelen, and W. W. C. Gieskes, *Mar. Ecol. Prog. Ser.*, 1997, **153**, 91.
[63] W. W. C. Gieskes and A. G. J. Buma, *Plant Ecol.*, 1997, **128**, 16.

Figure 6 Comparison of chemical action spectra and biological weighting functions. Solid line (left Y axis): average chemical action spectrum for production of hydrogen peroxide at 25°C (product of quantum yield and absorbance coefficient, mol J^{-1} h^{-1} m^{-1}) based on spectra ($n = 8$) from the Southern, Atlantic, and Pacific Oceans, Chesapeake Bay, and Boothbay Harbor; shaded area represents the standard deviation of log ($\varepsilon(\lambda)$). Dashed line/symbols (right Y axis) average BWF for inhibition of photosynthesis (see previous figure) with the line indicating the best fit exponential function ($\varepsilon(\lambda) = \exp(m_0 + m_1\lambda)$). Note the slope ($m_1$) is very similar to the slope of the peroxide chemical action spectrum

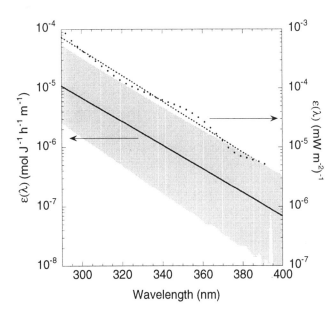

transformations occurring upon absorption of UV photons. Moreover, phytoplankton cells are the main source for marine dissolved organic matter in seawater.[64] It is reasonable, therefore, to expect that commonality to be reflected in the spectral structure of chemical action spectra and BWFs. Indeed, the spectral slope is quite similar for the average action spectrum of hydrogen peroxide production and the average BWF for inhibition of photosynthesis in natural phytoplankton assemblages (Figure 6). A log-linear approximation to the inhibition BWF has a spectral slope of 0.049 nm^{-1} ($R^2 = 0.97$), which is very close to the average slope of the chemical action spectra (0.050 nm^{-1}; see Section 2). The average action spectrum for hydrogen peroxide production in marine waters has a lower slope (0.046 nm^{-1}; Figure 6), because the spectrum is dominated by the production of estuary/bay waters which have somewhat shallower slopes than production in more pelagic samples. The quantum yield for hydrogen peroxide production has a similar spectral slope to the quantum yield for other photochemical processes.[17] Thus, the agreement in slopes between chemical action spectra and BWFs is indirect evidence that photolytic reactions also play a role in intercellular UV damage, especially in the UV-A band. For example, reactive oxygen species production by photosensitizing molecules is thought to be an important mode of cellular damage during UV exposure.[33] However, phytoplankton are generally unaffected by variations in the extracellular hydrogen peroxide and hydroxyl radicals, at least for environmentally plausible concentrations.[65] Finally, BWFs for inhibition of photosynthesis in cultures and for UV effects on organismal survival generally have a much steeper spectral

[64] C. Lee and S. G. Wakeham, *Mar. Chem.*, 1992, **39**, 95.
[65] Y. Zhu, K. Mopper, D. J. Kieber, W. H. Jeffrey, J. Kase, P. J. Neale, J. J. Fritz, and R. F. Davis, presented at the 2000 Ocean Sciences Meeting, San Antonio, TX, 2000, OS37.

slope than chemical action spectra, especially in the UV-B (Figs. 4, 5). Overall, the biological effects of UV are best explained by a combination of direct molecular changes (*e.g.* thymidine dimers) and indirect effects through reactive oxygen species.[33]

5 Assessment of UV Effects

The preceding sections have shown that much progress has been made in defining chemical action spectra and BWFs for marine processes. While there are still a number of significant issues to be resolved in their estimation and application, it is nevertheless instructive to review the use of these functions in predicting the impact of UV in the marine environment. In Table 1, we summarize the range of responses to solar irradiance predicted by a set of some 50 BWFs determined for various natural systems and phytoplankton cultures. The table is a modification of a previously published analysis, which compares different measures of UV radiation (unweighted and weighted).[9] In this case, we compare predicted responses to natural exposures as changes in photosynthesis or organismal survival. Example surface spectra are shown to illustrate how responses are influenced by typical environmental variation in UV. The applicability of these predictions vary. Certainly, it is sensible to apply Antarctic BWFs to responses at 60°S or 78°S, but there is little ecological relevance to using these same BWFs to predict responses at the equator. Moreover, the bulk of phytoplankton biomass is below the surface layer in much of the tropical ocean.

A primary determinant of UV effectiveness is sun angle, which is revealed (for constant ozone) in the variation of UV effects with latitude. On average, a 1 h exposure of noon surface irradiance at 45°N on the equinox (column 1, top) halves photosynthetic rates in natural populations. Exposure to equatorial intensities of UV induces a further inhibition so that rates are only around 70% of those at 45°N (column 2, bottom). The lower sensitivity, on average, of culture BWFs (Figure 4) has a significant effect on predicted response, the decrease at 45°N is only about 30%, and more than 80% of this rate is maintained even under equatorial exposure. Though the average response is remarkably similar between environments, there is considerable spatiotemporal variability in the predicted effects of UV within all BWFs measured in each environment. Typically, there is two- to three-fold variation in response between assemblages with the most and least sensitivity. Clearly, the variation in BWFs that has been encountered in measurements to date can have a strong impact on predicting responses to environmental UV at any one time or place.

Stratospheric ozone is a significant factor influencing incident UV-B, though effects are secondary after solar elevation and cloud cover (Table 1). The BWFs that have been determined to date suggest a moderate sensitivity to the severe depletion (>50%) that accompanies the polar ozone hole (Table 1, columns 9 and 10), with some 10–15% lower photosynthesis at surface intensities. As for overall response to UV, BWFs determined for cultures predict less additional inhibition due to ozone depletion enhanced UV-B. On the other hand, survival of zooplankton is strongly responsive to ozone depletion, as would be expected

Table 1 Predicted effects of UV exposure using biological weighting functions[a]

	1	2	3	4	5	6	7	8	9	10
					60°S O₃		Measured	Measured	% Effect of O₃ hole	
	45°N noon	Equator noon	Model 60°S normal O₃	Model 60°S O₃ hole	hole + 40% cloud reduction	60°S low O₃ 5 m	McMurdo normal O₃	McMurdo O₃ hole	60°S	McMurdo
WSC	43 (23–63)	28 (13–45)	44 (24–65)	38 (20–57)	52 (32–70)	75 (57–90)	42 (23–63)	36 (19–54)	14 (9–17)	15 (10–17)
Palmer	41 (30–66)	29 (19–52)	43 (31–67)	38 (27–62)	50 (38–73)	71 (59–87)	40 (29–66)	36 (25–60)	11 (7–16)	12 (7–17)
Antarctic average	43 (23–69)	30 (9–39)	44 (25–72)	40 (19–62)	52 (39–94)	73 (96–152)	42 (22–66)	37 (16–56)	12 (3–17)	13 (3–17)
Rhode River	43 (20–64)	30 (8–34)	44 (22–68)	39 (16–53)	51 (31–83)	71 (76–146)	42 (20–65)	37 (14–50)	13 (6–17)	13 (6–17)
Cultures	68 (56–90)	56 (34–67)	69 (58–92)	64 (48–82)	74 (69–100)	86 (101–121)	68 (56–90)	64 (47–83)	8 (3–14)	7 (2–12)
Copepod survival	66	41	70	47	64	88	73	55	33	24
Relative to 45°N										
WSC	1.00	0.64	1.04	0.89	1.23	1.74	0.99	0.84		
Palmer	1.00	0.70	1.03	0.92	1.21	1.72	0.98	0.87		
Antarctic average	1.00	0.69	1.03	0.92	1.21	1.69	0.98	0.87		
Rhode River	1.00	0.69	1.03	0.91	1.19	1.66	0.99	0.87		
Cultures	1.00	0.82	1.02	0.94	1.09	1.27	1.00	0.94		
Copepod survival	1.00	0.61	1.72	0.67	1.35	1.38	0.83	0.76		

[a] The mid-day spectra are as previously described;[9] mid-day spectral irradiance at 45°N and the equator are simulated for 300 Dobson Units (DU: 10^{-3} cm O_3 at STP); spectra for 60°S are for high O_3 (340 DU) and low O_3 (140 DU) during the austral spring. Effects of clouds on low O_3 spectra are approximated with a 40% reduction at all wavelengths. Irradiance at 5 m is estimated using diffuse attenuation coefficients from the Weddell–Scotia Confluence.[66] Measured spectra at McMurdo Station, Antarctica (78°S) on 28 October 1990 (ozone hole, 175 DU) and 10 November 1990 (normal, 350 DU). To correct for cloud cover, the spectrum on 28 October was multiplied by a factor of 1.53 to match that of 10 November for integrated irradiance, 350–400 nm. The individual BWFs summarized in Figures 4–6 were applied to these spectra and percent relative response (average photosynthetic rate or probability of survival) was calculated for a 1 h exposure using the appropriate ERC. The mean response (followed by minimum to maximum range) is given in the upper part of the table; the corresponding row in the lower panel is each measure normalized to that modeled for 45°N. The inhibition due to O_3 depletion is calculated from the reduction in photosynthesis between high and low ozone column spectra at 60°S and McMurdo, i.e. $(P_{HI} - P_{LO})/P_{HI}$ where P is photosynthesis or probability of survival at normal (HI) or reduced (LO) ozone.

Table 2 Predicted effects of UV exposure using photochemical action spectra[a]

	45°N noon	Equator noon	Model 60°S normal O_3	Model 60°S O_3 hole	60°S O_3 hole + 40% cloud reduction	60°S low O_3 5 m	Measured McMurdo normal O_3	Measured McMurdo O_3 hole	Ozone hole enhancement 60°S	McMurdo
Boothbay Harbor	105	185	99	118	71	31	107	130	19%	21%
Chesapeake	75	136	71	90	54	22	75	96	27%	28%
Banks Channel, NC	59	104	56	66	40	17	60	73	19%	21%
Mean Estuary/Bay	80	141	75	91	55	23	81	100	22%	23%
Gulf of Maine Sta. B Hawaii station	8.0	14.1	7.6	9.0	5.4	2.4	8.2	9.9	18%	20%
ALOHA	1.7	3.2	1.6	2.2	1.3	0.5	1.7	2.2	36%	35%
Mean temp. ocean	4.9	8.6	4.6	5.6	3.4	1.5	4.9	6.1	27%	28%
WSC Station N	0.42	0.77	0.40	0.51	0.31	0.12	0.42	0.54	28%	28%
WSC Station B	0.81	1.48	0.75	0.99	0.60	0.23	0.79	1.04	32%	32%
Palmer Station	1.34	2.43	1.26	1.61	0.96	0.39	1.33	1.71	28%	28%
Mean Antarctica	0.86	1.56	0.80	1.04	0.62	0.25	0.85	1.10	29%	30%
Average all areas	24.4	43.6	23.1	28.3	17.0	7.2	24.7	30.7	22%	24%
Relative to 45°N										
Average production	1.00	1.78	0.95	1.16	0.69	0.29	1.01	1.26		
Reciprocal production	1.00	0.56	1.06	0.86	1.44	3.41	0.99	0.80		

[a] The midday spectra are as for Table 1. The chemical action spectra are the means for samples from the individual study areas as specified, the same chemical action spectra as summarized in Figure 6. All rates (μmol H_2O_2 m^{-3} h^{-1}) were calculated for surface exposure at nominal sample temperature, 25°C for temperate areas and 0°C for Antarctica. The bottom block gives the average production normalized to that modeled for 45°N. The relative response was within 5% of this proportion for all of the individual chemical action spectra. Also given is the reciprocal relative production for comparison with relative biological response in Table 1. The enhancement due to O_3 depletion is calculated from the increase in photochemistry between high and low ozone column spectra at 60°S and McMurdo, i.e. $(R_{LO} - R_{HI})/R_{HI}$ where R is the rate at normal (HI) or reduced (LO) ozone.

given that BWFs mainly weight the UV-B band (*cf.* Figure 5).

Predicted rates of UV-induced photochemistry using chemical action spectra are illustrated in Table 2, using the same example irradiance spectra as for Table 1. These are surface rates (optically thin) for photochemical production of dissolved hydrogen peroxide (H_2O_2), but the relative response is similar for many other processes. Some action spectra that diverge significantly from these responses have been previously described.[12] Predicted absolute rates of production at the surface vary by about 100–fold, which mainly reflects the difference in absorbance between CDOM-rich estuarine waters and CDOM-poor oceanic regions (including Antarctica). Also, quantum yields decrease with temperature.[18] The predicted rates agree fairly well with observed rates in temperate waters, but somewhat underestimate measurements in Antarctic waters.[18] Some of the discrepancy is probably due to the omission of the minor component of production (10–15%) due to visible radiation. Relative rates of H_2O_2 production between irradiance spectra vary less than 5%, so only a single average proportion is given. The relative variation in the reciprocal of the production rate ($1/R_{H_2O_2}$) is also given to compare with the biological responses given in Table 1. This is based on the assumption that loss of biological function should vary inversely with rate of photochemistry. Actually, the variation in photochemistry is greater than the variation in biological response. This might seem in conflict with the general agreement between chemical action spectra and BWFs shown in Figure 6. However, the difference is due to nonlinear ERCs for biological responses compared to an assumed linear relationship between irradiance and photochemical production rate. Also, ozone depletion leads to a greater relative increase in photochemical production than biological damage. Again, this is a consequence of the nonlinearity of biological response over a 1 h exposure.

These calculations address the simple question of what happens during a 1 h exposure at noon. It is a straightforward, though ambitious, undertaking to extend the calculation over time, depth, and space. An additional complication in evaluating impacts in aquatic systems is accounting for the changes in duration of exposure arising from vertical mixing. This was accomplished by integrating a numerical model of vertical mixing with the BWF_H-PI model to predict UV effects on daily phytoplankton productivity in the WSC.[66] The results suggest that absolute effects are much less when integrated daily primary productivity (PPR, $gC\,m^{-2}\,d^{-1}$) is considered; however, the trends of the responses are similar to the trends in Table 1. Differences in WSC BWFs led to an approximate factor of three variation in the effect of full spectral UV on PPR, and an approximate factor of two variation in the effect of ozone depletion. Averaged over mixing conditions ranging from static to 100 m mixed layer, and 20 min to 8 h mixing times, a 50% drop in ozone depletion led to 1.5–3.5% further decrease in water column PPR. The response to ozone depletion for the worst combination of BWF, mixing depth, and mixing time was a 8.5% decline in PPR. Another model using different assumptions (in particular, no vertical mixing) has predicted less impact from ozone depletion on Southern Ocean productivity.[67]

Photochemical production models are now being developed that consider the

[66] P. J. Neale, R. F. Davis, and J. J. Cullen, *Nature*, 1998, **392**, 585.
[67] K. R. Arrigo, *Mar. Ecol. Prog. Ser.*, 1994, **114**, 1.

effect of mixing on production rates. When mixing was considered, measured and predicted rates of carbon monoxide photoproduction agreed extremely well.[68] A numerical modeling approach is also being used to examine the interaction of mixing and the damage and repair of DNA in bacteria[69] and survival of zooplankton and fish eggs.[70] Results for the UV-sensitive copepod, *C. finmarchicus*, were similar to those of phytoplankton: shallow mixing tend to enhance UV effects, whereas deep mixing tended to counteract the effects. This result is not too surprising, considering that overall sensitivity of the processes examined are similar (Figure 5); however, the specific interactions between vertical mixing and response differ. For phytoplankton, deep mixing enhances photosynthesis; for zooplankton, it hastens transport of negatively buoyant eggs out of the surface layer.

Models for the impact of UV on marine biological processes have so far shown only mild, though significant, effects of ozone depletion. A strong dependence on ozone depletion was observed in modeling of water-column DNA damage (thymidine dimers) in the environment,[69] but this type of DNA damage is rapidly repaired and thus may not be a good indicator of overall organismal response. However, ozone depletion is not the only way that global change contributes to variations in environmental UV. Changes in CDOM will also affect the penetration of UV into natural waters, and such variations can be equal or greater than variations linked to ozone depletion.[71]

6 Summary and Conclusions

The informed application of basic photochemical and photobiological principles to marine environments is improving our understanding of the effects of solar ultraviolet radiation. A number of successful techniques have been developed to generate data for responses to UV under environmentally relevant conditions. There are some areas where more work is needed, for example better absorption measurements of CDOM and BWFs for more processes than phytoplankton photosynthesis. Quantitative models which include spectral weighting functions are being developed for an increasing number of marine processes and the result is the clarification of the role of UV in the environment, at least over a limited range of time and space scales (up to kms and days). The range of environments that have been studied and our understanding of general patterns in responses should increase in the next few years.

In the future, it should be possible to increase the temporal and spatial scale over which realistic modeling analysis can be attempted. Our review of weighting functions determined to date shows significant variability in responses, yet also a striking similarity in responses between environments. It will be interesting to see if this pattern continues to hold as more assemblages and species are studied. On the other hand, detailed comparative studies should lead to a better understanding

[68] R. G. Najjar, J. Werner, O. C. Zafiriou, H. Xie, W. Wang, and C. D. Taylor, presented at the 2000 Ocean Sciences Meeting, San Antonio, TX, 2000, OS93.
[69] Y. Huot, W. H. Jeffrey, R. F. Davis, and J. J. Cullen, *Photochem. Photobiol.*, 2000, **72**, 62.
[70] P. S. Kuhn, H. I. Browman, R. F. Davis, J. J. Cullen, and B. McArthur, *Limnol. Oceanogr.*, 2000, submitted.
[71] D. W. Schindler, P. J. Curtis, B. Parker, and M. P. Stainton, *Nature*, 1996, **379**, 705.

of what factors determine sensitivity and differential responses with wavelength. As a systematic understanding of the variations in chemical action spectra and BWFs is developed, the variability in responses can be incorporated in modeling efforts. Whereas there is considerable uncertainty about the scenario for the recovery of the ozone layer over the next century, we should have a better idea of the chemical and biological impact of whatever UV conditions occur.

Effects of Solar UV-B Radiation on Terrestrial Biota

JELTE ROZEMA

1 Evolution of Terrestrial Biota and the Stratospheric Ozone Layer

Since the dramatic discovery of the ozone hole over the Antarctic in 1985, being caused by the release of CFCs, there is great concern about this ozone depletion and the consequences of stratospheric ozone breakdown. Yet it is also important to know about UV-B radiation levels in past times. Has solar UV-B radiation been higher in the past? How did plants respond to earlier UV-B radiation levels? Have adaptations of plants to past solar UV-B fluxes been maintained or lost? Here we briefly review the evolution of the atmosphere, and the evolution of the stratospheric ozone layer in some more detail.

There is strong evidence that in previous times UV-B radiation on Earth was much higher than at present. In the early history of the Earth, photosynthesis of marine and fresh water aquatic autotrophic organisms eventually led to an increase of atmospheric oxygen and a decrease of atmospheric carbon dioxide (Figure 1), which made possible the development of a stratospheric ozone layer. In early times, however, no plant life on land was possible because of too high UV-B fluxes. High levels of solar UV-B radiation at the Earth's surface in the reducing primeval atmosphere forced early organisms to live submerged in the water. Attenuation of solar UV-B radiation by the water column is assumed to have created an environment sufficiently low in solar UV-B radiation to allow photosynthesis and growth of prokaryotic organisms. Thus, early submerged aquatic organisms used the water column as an external UV screen.

The young Earth, about 4×10^9 years ago, received very high doses of solar UV radiation. It is estimated that, at that time, the sun emitted about 10 000 times more UV than at present. There was liquid water, because high atmospheric carbon dioxide levels, about 100–1000 times higher than present values, absorbed infrared radiation and created a greenhouse effect. There was no oxygen in the atmosphere before the occurrence of oxygenic photosynthesis. There is only indirect evidence for the occurrence of oxygen in the atmosphere. Deposition of iron oxide in Red Beds, dated 2.0×10^9 years ago, indicates aerobic terrestrial weathering. Atmospheric oxygen at that time is estimated to have amounted to 0.001% of the present level. With the development of oxygenic photosynthesis, 2.7×10^9 years before present, oxygen in the atmosphere

Figure 1 Relative concentrations of atmospheric oxygen (O_2) and carbon dioxide (CO_2) during the evolutionary history of the Earth, development of the stratospheric ozone layer and the advent of higher plants as part of the terrestrial biota. Atmospheric oxygen is derived from photosynthetic activity of autotrophic organisms during the photolysis of water molecules. At the same time, carbon dioxide is being fixed to a CO_2 receptor to form carbohydrates

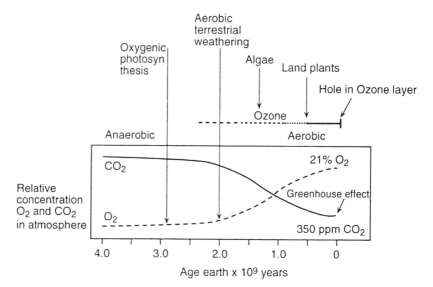

increased. The development of stratospheric ozone might have been in proportion to the gradual atmospheric oxygen increase. Alternatively, a steep rise of atmospheric oxygen to near present levels of 21% may have been reached 350×10^6 years ago.[1,2] In any case, stratospheric ozone, which absorbs solar UV-C completely and much of the UV-B radiation, reduced the flux of damaging solar UV on the Earth's surface and must have allowed the evolution of terrestrial plant life. The origin of the earliest land plants has been dated 470×10^6 years ago. It is generally assumed that the build up of stratospheric ozone, tracking atmospheric oxygen levels, has greatly influenced the timing of land plant evolution. Continuous release of oxygen by photosynthesis of photosynthetic aquatic bacteria, cyanobacteria, eukaryotic algae and eventually terrestrial plants has led to a gradual increase of atmospheric oxygen and a concomitant decrease of carbon dioxide in the atmosphere (Figure 1).

There is no general agreement on the question of the timing of the occurrence of oxygen in the atmosphere and thus the development of the stratospheric ozone layer, since the reconstruction of historical levels of atmospheric oxygen remains indirect and uncertain. There is evidence that, during the evolution of land plants, the thickness of the stratospheric ozone layer was less (and the flux of solar UV-B radiation higher) than at present.[3–5] One implication of this is that the recent increase of solar UV-B radiation is not as high as during the early evolution of land plants. It may be hypothesized that early land plants adapted to such high UV-B fluxes. This might imply that some 'old' terrestrial plant groups, *e.g.*

[1] R. A. Berner, *Paleogeogr. Paleoclimatol., Palaeoecol.*, 1989, **75**, 97.
[2] R. A. Berner, *Science*, 1993, **261**, 68.
[3] L. Margulis, J. C. G. Walker and M. Rambler, *Nature*, 1976, **264**, 620.
[4] M. M. Caldwell, *Bioscience*, 1979, **29**, 520.
[5] J. Rozema, J. W. M. van de Staaij, L. O. Björn and M. M. C. Caldwell, *UV-B as an environmental factor in plant life: stress and regulation. Trends Ecol. Evol.* 1997, **12**, 22–28.

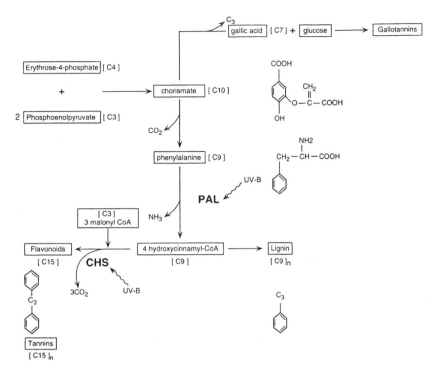

Figure 2 Induction of PAL and CHS of the phenylpropanoid pathway by (enhanced) solar UV-B. C3 and C4 compounds of the 'Primary metabolism' combine to form C10 chorismate. In the shikimic acid pathway the aromatic amino acid phenylalanine (C9) is formed after decarboxylation. The UV-B inducible enzyme PAL catalyzes a deamination reaction. Lignin is a complex phenolic macromolecule based on C9 units. CHS catalyses the formation of C15 phenolic units such as flavonoids. Tannins represent polyphenolics based on such C15 phenolic units

terrestrial algae and lichens, (may) have become well adapted to such high UV-B radiation periods.

The development of stratospheric ozone, as a result of the rise of oxygen in the atmosphere, will have influenced the timing of the evolution of land plants. Without the ozone layer, shortwave UV-C and UV-B radiation likely prevented much evolution of terrestrial plant and animal life.[5] With the development of a stratospheric ozone shield (another external UV screen), much of the damaging solar UV was removed and this is thought to have allowed terrestrial plant life to evolve from aquatic environments. However, plants were also developing internal UV filters, largely through secondary metabolism and production of (poly)phenolics. The timing of these events and how they contributed to terrestrial plant evolution is still uncertain.

UV-B Radiation and the Evolution of Polyphenolics

Solar UV-B radiation is known to stimulate the enzymes phenylalanine ammonia lyase (PAL) and chalcone synthase (CHS) and other branch-point enzymes of the phenylpropanoid pathway (Figure 2).[6-8] PAL catalyzes the transformation of phenylalanine into *trans*-cinnamic acid, which may eventually

---

[6] C. J. Beggs and E. Wellmann, Photocontrol of flavonoid biosynthesis. *Photomorphogenesis in plants* (eds. R. E. Kendrick and G. H. M. Kronenberg), 1994, pp. 733–751. Kluwer Academic Publishers, Dordrecht.

[7] L. Liu, D. C. Gitz III and J. W. McClure, *Physiol. Plant.*, **93**, 725.

[8] G. A. Cooper-Driver and M. Bhattacharya, *Phytochemistry*, 1998, **49**, 1165.

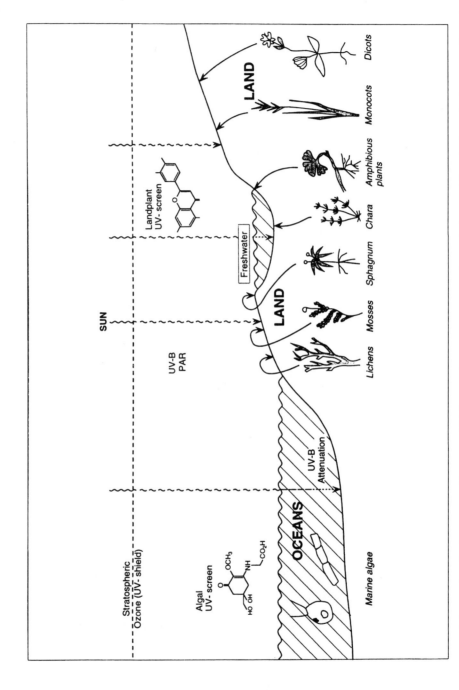

Figure 3 Survey of the marine, fresh water and lower and higher land plant groups and the evolution and functioning of UV screens in various water and land plants. Algae are assumed to have relatively simple phenolics and chemically more complex polyphenolics as UV screens are expected to occur more frequently in higher plants

Figure 4 Structural formulae of UV-B-absorbing compounds in marine algae (*e.g. Ulothrix*), lower land plants, *e.g.* Sphagnaceae (*e.g. Sphagnum*), mosses (*e.g. Hylocomium*) and higher land plants (*e.g. Silene vulgaris*). An increasing chemical complexity from lower to higher plants evolving from the water to the land is indicated

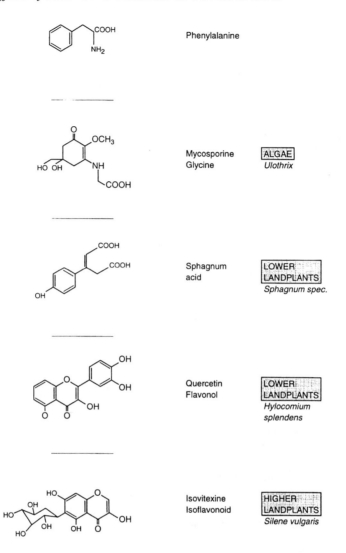

lead to the formation of complex phenolic compounds such as flavonoids, tannins and lignin (Figure 2).

UV absorption in the UV-B wavelengths by cinnamic acid (maximum absorption at 308 nm) exceeds that of phenylalanine (maximum absorption at 280 nm). The chemical evolution of these and other UV-B-absorbing compounds has been assumed to be an important part of the evolution from aquatic plants to land plants.[9,10] The complexity of polyphenolics tends to increase with evolutionary advancement from algae, charophycean algae, bryophytes, pteridophytes, gymnosperms to angiosperms (Figures 3 and 4).[1,11] A high degree

[9] K. Kubitzki, *J. Plant Physiol.*, 1987, **131**, 17.
[10] H. E. Stafford, *Plant Physiol.*, 1991, **96**, 680.
[11] J. M. Robinson, *Geology*, 1980, **15**, 607.

of polymerization of phenolics is found in land plants, *i.e.* flavonoids, tannins and lignin.[5,8]

In addition to absorption of UV-B radiation, these compounds serve many other functions such as signal transduction, plant hormones, chemical defence against microorganisms and herbivory, and structural rigidity (Figure 5). This structural rigidity of lignified higher plants serves the water transport through xylem vessels. The upright growth form of such land plants, for example, allows acquisition of sunlight and outcompeting of plants with less tall growth forms.

The UV-B-induced (poly)phenolics in plants, formed in the phenylpropanoid pathway (Figure 2), do not just serve as plant protection to UV-B radiation and physiological and ecological processes. Polyphenolics such as lignin and sporopollenin tend to be physically and chemically stable and may remain to a large extent intact after the death of a plant.

2 Solar UV-B, Polyphenolics, the Pool of Organic Carbon in Terrestrial Environments, and the Balance between Oxygen and Carbon Dioxide in the Earth's Atmosphere

The physical and chemical stability and the resistance to microbial breakdown of complex polyphenolics in plant litter contributes considerably to the long-term storage of organic carbon in terrestrial ecosystems.[1,2,5] The polyphenolic compound lignin determines to a large extent the resistance of dead organic plant matter to degradation by microorganisms. Lignin contributes to the persistence of peat and humus in soils. Thereby, complex polyphenolics help to establish large, long-lasting sinks of carbon in terrestrial ecosystems. The established long-term relative homeostasis between oxygen and carbon dioxide in the current atmosphere[1,2,5] is partly based on the presence of recalcitrant polyphenolics in terrestrial soil biota.[8,12,13]

In some recent studies,[14-17] enhanced solar UV-B radiation, simulating about 15% stratospheric ozone depletion, resulted in an increased content of tannins and lignin in plant tissue in terrestrial plants. Decomposition of such plant litter with increased content of tannins and lignin was retarded compared to litter from plants which were not grown under enhanced UV-B radiation.

3 Current Stratospheric Ozone Depletion: Increased Solar UV-B Radiation Reaching the Earth

Depletion of stratospheric ozone due to the release of CFCs and nitrous oxides was first[18] reported for the Antarctic in 1985 and now the so-called Antarctic

[12] L. E. Graham, *Origin of Land Plants*. Wiley, New York, 1993.
[13] J. Rozema, B. Kooi, R. Broekman and L. Kuijper, in *Stratospheric Ozone Depletion: Effects of Enhanced UV-B on Terrestrial Ecosystems*, ed. J. Rozema, Backhuys, Leiden, 1999, pp. 117–134.
[14] C. Gehrke, U. Johansson, D. Gwynn-Jones, L.-O. Björn, T. V. Callaghan and J. A. Lee, *Ecol. Bull.*, 1996, **45**, 192.
[15] J. Rozema, M. Tosserams, H. J. M. Nelissen, L. van Heerwaarden, R. A. Broekman and N. Flierman, *Plant Ecol.*, 1997, **128**, 284.
[16] C. Gehrke, U. Johansson, T. V. Callaghan, D. Chadwick and C. H. Robinson, *Oikos*, 1995, **72**, 213.
[17] D. Gwynn-Jones, J. A. Lee, T. V. Callaghan and M. Sonesson, *Plant Ecol.*, 1997, **128**, 242.
[18] L. O. Björn, *J. Environ. Sci.*, 1996, **51**, 217.

Effects of Solar UV-B Radiation on Terrestrial Biota

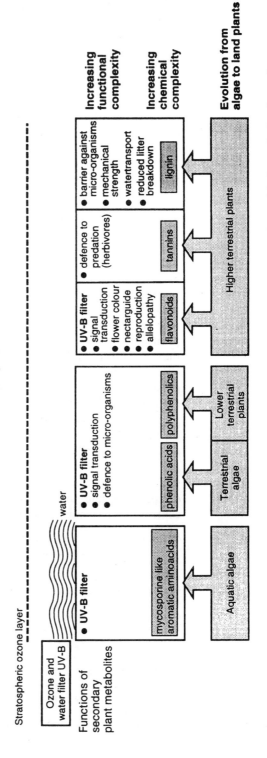

Figure 5 Increasing chemical complexity and functional complexity of (poly)phenolic compounds along with the evolution from algae to land plants

ozone hole exists, particularly during the Antarctic spring.[18-20] There is now extensive scientific interest in stratospheric ozone depletion and the consequences of enhanced solar UV-B radiation for terrestrial biota.[19-21] Low stratospheric ozone values occur over the Antarctic, and enhanced solar UV-B radiation as a result of stratospheric ozone depletion may cause damage to plants, although UV-B protective mechanisms appear as well.[5,18,22] UV-B damage such as the occurrence of thymine–thymine dimers in DNA may be repaired by the enzyme DNA photolyase.[18] The activity of this enzyme is temperature limited, however, and Arctic as well as Antarctic plants may therefore be sensitive to enhanced UV-B because of low Antarctic temperatures. Plants may further protect themselves to UV-B damage by morphogenic changes and the development of UV-B-absorbing screens.[23-26]

4 Effects of Enhanced Solar UV-B Radiation on Terrestrial Plants, Adaptations of Terrestrial Plants to Solar UV-B: Evidence from Physiological Studies

Several reports indicate that ambient levels of solar UV-B represent environmental stress to plants. This implies that plant growth is reduced by ambient solar UV-B radiation, compared with below-ambient solar UV-B. This environmental stress may be transient. For example, reduction in the length of developing leaves of plant species from a dune grassland at ambient UV-B disappeared later during the growing season. Much of the earlier—(eco)physiological—research of UV-B effects on plants focused on damage caused by UV-B radiation (Table 1). Of the possible primary metabolic targets of UV-B radiation to plants, DNA damage is currently considered to be most important (Figure 6).

Since UV-B radiation may cause damage to membranes, DNA,[18] and various other cellular structures and processes, it was expected that plant and crop growth would suffer from enhanced solar UV-B radiation.[5,18,22] Responses of plant species from various latitudes and from low and high UV-B environments to UV-B have been reported. Many of these experimental studies of UV-B effects on plants relate to crop plants exposed to enhanced UV-B in indoor controlled-environment experiments.[24,27-29] In relatively few studies have

[19] J. Rozema (ed.), *Stratospheric Ozone Depletion: the Effects of Enhanced UV-B on Terrestrial Ecosystems*, Backhuys, Leiden, 1999.
[20] M. M. Caldwell, L.-O. Bjorn, J. F. Bornman, S. D. Flint, G. Kulandaivelu, A. H. Teramura and M. Tevini, *J. Photochem. Photobiol. B*, 1998, **46**, 40.
[21] M. M. Caldwell and S. D. Flint, *Climatic Change*, 1994, **28**, 375.
[22] M. M. Caldwell, A. H. Teramura and M. Tevini, *Trends Ecol. Evol.*, 1989, **4**, 363.
[23] M. Tosserams and J. Rozema, *Environ. Pollut.*, 1995, **89**, 209.
[24] M. A. K. Jansen, V. Gaba and B. M. Greenberg, *Trends Plant Sci.*, 1998, **3**, 131.
[25] T. A. Day, T. C. Vogelmann and E. H. DeLucia, *Oecologia*, 1992, **92**, 513.
[26] B. Meijkamp, R. Aerts, J. van de Staaij, M. Tosserams, W. H. O. Ernst and J. Rozema, in *Stratospheric Ozone Depletion: Effects of Enhanced UV-B on Terrestrial Ecosystems*, ed. J. Rozema, Backhuys, Leiden, 1999, pp. 71–100.
[27] A. R. McLeod, *Plant Ecol.*, 1997, **128**, 78.
[28] E. L. Fiscus and F. L. Booker, *Photosynth. Res.*, 1995, **43**, 81.
[29] J. Rozema, J. W. M. van de Staaij and M. Tosserams, in *Plants and UV-B Responses to Environmental Change*, ed. P. J. Lumsden, Cambridge University Press, Cambridge, 1997, pp. 213–232.

Table 1 Potential direct and indirect effects of enhanced UV-B radiation on terrestrial plant growth and processes in terrestrial ecosystems

Direct effects
1. DNA damage
    Cyclobutane dimer formation
    (6–4) photoproduct formation
2. Photosynthesis
    Disturbance of Photosystem II
    Thylakoid membrane functioning
    Stomatal functioning
3. Membrane functioning
    Peroxidation of unsaturated fatty acids
    Damage to membrane proteins
4. Stimulation of PAL, TAL and CHS of phenylpropanoid pathway

Indirect effects
1. Plantmorphogenetic effects
    Leaf thickness
    Leaf angle
    Plant architecture
    Biomass allocation
2. Plant phenology
    Emergence
    Senescence
    Flowering
    Reproduction
3. Chemical composition of plants
    Tannins
    Lignin
    Flavonoids

responses of plants from natural ecosystems to enhanced UV-B been reported.[5,19,20] So far, no marked direct effects of enhanced UV-B in lowering primary production of terrestrial ecosystems have been found.[19,20] Terrestrial plants absorb PAR wavelengths to drive photosynthestic processes in the chloroplast. This implies that there must be sufficient transmission of PAR through the outer cuticular and epidermal leaf tissues. Leaves of higher plants also absorb a large fraction of incident solar UV-B radiation, mainly by flavonoids and related phenolics located primarily in the epidermis. Cuticular waxes are not strong absorbers of UV-B. In some species, (poly)phenolics form part of cuticular leaf hairs or bladders and these structures can be important UV-B absorbers. Generally, most of the incident UV-B is absorbed by leaves, a minor fraction is reflected or scattered at the leaf surface, and only a negligible amount can be transmitted through the entire leaf.[30]

Damage by UV-B to key targets in plants can be largely avoided by effective, wavelength-selective filtering of UV-B by phenolics localized in epidermal cells (Table 2). Other defences against UV-B damage consist of (1) DNA repair, (2) scavenging of radicals resulting from absorption of UV-B photons, and (3) the production of polyamines, which stabilize injured membranes.

In a recent study,[31] Rozema et al. found reduced shoot length, increased leaf

---

[30] A. J. Visser, M. Tosserams, M. W. Groen, G. Kalis, R. Kwant, G. W. H. Magendans and J. Rozema, *Plant Ecol.*, 1997, **28**, 208.

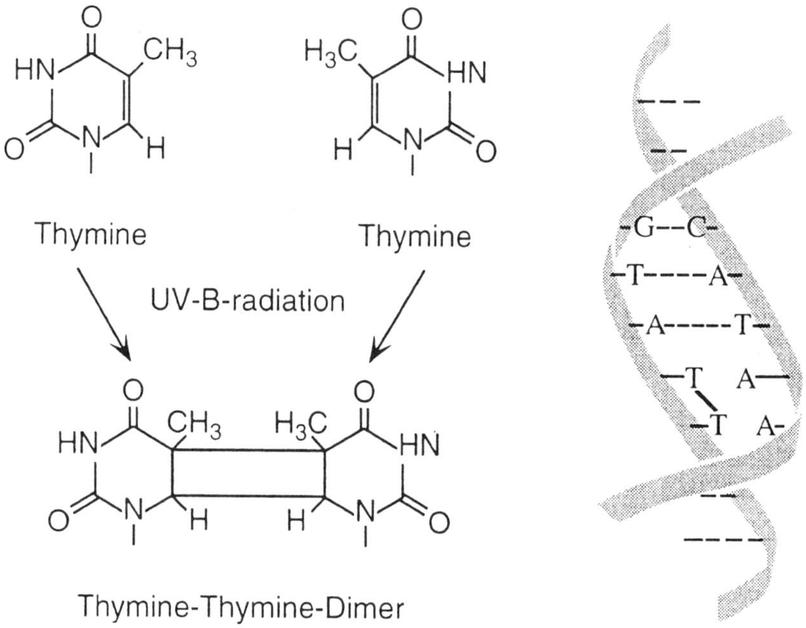

Figure 6 Absorption of UV-B photons may lead to disruption of the hydrogen bonds between the base pairs thymine and adenine. As a result, cyclobutane dimers of thymine groups may be formed. In the light the enzyme DNA-photolyase cleaves these dimers and the coupling of thymine and adenine is restored (photorepair). In the Arctic and antarctic, low temperatures may lead to a low activity of DNA-photolyase

Table 2 Adaptation mechanisms of higher plants against damage by enhanced UV-B radiation

1. Scattering and reflection of UV-B radiation by epidermal and cuticular structures; other leaf optical properties such as wax layer and leaf hairs, leaf bladders
2. Absorption of UV-B radiation by pigments (flavonoids, anthocyanins), particularly in the epidermal cells. Most of the intercepted UV-B is absorbed in this way by plant leaves
3. Photoreactivation enzymes (photolyases): monomerization of dimers formed by DNA absorption of UV-B photons (photorepair). Photoreactivation is a rapid process, but needs sufficient PAR. Photorepair is regarded more important than excision repair
4. Excision repair of DNA damage caused by UV-B radiation. A slow process, occurring also in the dark
5. Scavenging of radicals formed by absorption of UV-B photons by superoxide dismutase (SOD) and catalase. Flavonoids may also be involved in neutralizing radicals
6. Polyamines may ameliorate UV-B damage to membranes

thickness and increased branching, and more erectophyllous leaves in *Deschampsia antarctica* exposed to enhanced UV-B radiation (Figure 7). The RGR (relative growth rate) was not affected by UV-B, probably because increased branching compensated for decreased leaf and shoot length. Net leaf photosynthesis per unit area was not significantly affected by elevated UV-B radiation. It was concluded that reduced plant height increased branching and more but smaller erectophyllous leaves was an acclimation response to UV-B and served to reduce absorbance of high UV-B levels.

[31] SORG (Stratospheric Ozone Review Group), *Stratospheric Ozone* 1999, The Stationery Office, London, 1999.

Figure 7 The response of Antarctic hairgrass, *Deschampsia antarctica*, to enhance UV-B. In this climate room experiment with 0, 2.5 and 5 kJ m^{-2} day^{-1} UV-B$_{BE}$ treatment, the length in growth of shoots at 5 kJ m^{-2} day^{-1} UV-B$_{BE}$ was markedly reduced compared to 0 and 2.5 kJ m^{-2} day^{-1} UV-B$_{BE}$. In addition, there was an increased number of shoots and increased leaf thickness with enhanced UV-B. The relative growth rate (RGR) was not affected by UV-B, possibly because reduced shoot length growth by enhanced UV-B was compensated by increased tillering. Light response curves of net leaf photosynthesis of plants exposed to 5 kJ m^{-2} day^{-1} UV-B$_{BE}$ did not differ from those exposed to 0 kJ m^{-2} day^{-1} UV-B$_{BE}$. The content of UV-B-absorbing compounds of plants exposed to increasing UV-B did not significantly change

The above example clearly demonstrates the interaction between indirect photomorphogenic or growth responses and that of direct effects of UV-B on photosynthesis *per se*. It also points to the complexity of the whole plant response to UV-B radiation and the need for further understanding of the mechanisms of both the photomorphogenic and the photosynthetic responses.

5 Methodologies for the Study of UV-B Effects on Plants of Terrestrial Biota

In assessing the effects of enhanced solar UV-B radiation on terrestrial plants, different experimental methods are followed (Table 3). Until recently, most experimental UV-B studies were done in climate rooms and greenhouses: so-called controlled-environment studies.[21,27] In general, the scientific benefits of such an approach are clear. Environmental conditions such as light, temperature and humidity can be controlled and differences in plant growth can be ascribed to only one experimental factor, say nutrient supply. In the case of studies of the effects of enhanced UV-B radiation in controlled environments to simulate stratospheric ozone depletion, it has appeared to be difficult to mimic the outdoor spectral solar radiation. First, PAR levels in controlled-environment studies are often much lower than in the field. Increased indoor PAR levels are often associated with increased heat release and unwanted temperature rise. Second, indoor light is usually supplied by lamps of which the spectral radiation differs from that of solar radiation.

Because of low indoor PAR/UV-B ratios and differences between PAR and the UV-lamp spectra and that of solar radiation, the results of such indoor UV-B

Table 3 Survey of methods used to vary levels of UV-B radiation in indoor and outdoor experiments[a]

Indoor	1.	UV-B fluorescent tubes in combination with PAR light souces. Cut-off filters (*e.g.* cellulose acetate) to screen out UV-C and part of solar UV-B. Different levels of UV-B_{BE} by varying the period of exposure to UV-B and distance of UV tubes to plants. Widely used. Disadvantage: absence of spectral balance
Outdoor		Ambient – below ambient solar UV-B radiation
	2.	Filters (*e.g.* mylar foil) to reduce ambient solar UV-B radiation to below ambient levels. Disadvantage: precipitation is (partly) intercepted; temperature increases underneath filter. Advantage: requires no electricity; can be applied in remote areas
	3.	Ozone cuvette: special type of filter. The cuvette, with UV-B-transmitting glazing, is filled with ozone to reduce transmission of UV-B to a desired level. An advantage is a realistic spectral distribution. Only rarely used
		Ambient – above ambient solar UV-B radiation
	4A.	UV-B supplementation with UV fluorescent tubes Square wave mode. UV tubes, in combination with appropriate cut-off filters, are switched on and off generally around midday, to obtain an enhanced level of UV-B radiation above ambient solar UV-B radiation. Generally used
	4B.	Solar tracking mode (modulated irradiation system). Natural solar UV-B is continuously monitored and a constant proportional supplement of UV-B is maintained. Although in theory this is a better method of exposing plants to enhanced UV-B, the results of the application of the solar tracking UV-B enhancement mode do not differ from the square wave mode UV-B experiments

[a] For an overview of outdoor UV-B supplementation methods, see refs. 21 and 27.

effect studies cannot easily be extrapolated to outdoor conditions. As a result of the low indoor PAR/UV-B ratios and the absence of the natural solar spectral balance, the activity of the DNA-photolyase enzyme system may be limited. Therefore UV-B damage in indoor studies has been often overestimated. In outdoor conditions, much less significant UV-B effects have been reported.

In practice there are, at the moment, two ways of changing levels of UV-B radiation in the field (Table 3). By application of UV lamps in combination with appropriate filters, UV-B radiation can be enhanced (Figure 8). Based on actual outdoor UV-B radiation measurements or on model calculations and the use of a UV-B plant action spectrum, a weighted biologically effective UV-B dose can be obtained and varied, simulating the increase of UV-B radiation at the Earth's surface related to stratospheric ozone depletion (Figure 8).

Figure 9 shows a mini UV-B lamp system installed over *Deschampsia antarctica* at a north-faced slope of rocks at Léonie Island of the Antarctic peninsula. In the background is the polar ocean, with icebergs and glaciers at the horizon. Measurements have made clear that the burning mini-UV-B lamp systems homogeneously irradiates an area of about $40 \times 50 \text{cm}^2$ ground area.[19]

Alternatively, filters can be used to reduce levels of ambient, natural solar UV-B radiation levels. This will result in ambient solar UV-B and below ambient solar UV-B radiation levels. This will result in ambient solar UV-B and below ambient solar UV-B radiation levels (Figure 7). Such filters can also be used to reduce UV-B radiation emitted by UV lamps or tubes.

The advantage of filtration of natural solar UV-B radiation is that the natural balance between different wavebands (UV-B, UV-A, PAR) can be maintained. The disadvantage is that, at the moment, only ambient and below ambient UV-B radiation levels (and not above ambient UV-B radiation levels) can be obtained.

It is generally agreed that results obtained by studying the effects of different levels of natural solar UV-B radiation on terrestrial plants will be more realistic than those from indoor studies.[21]

6 Direct and Indirect UV-B Effects on Terrestrial Ecosystem Processes and Feedbacks, Autotrophic and Heterotrophic Relationships

Direct and Indirect Effects of UV-B on Terrestrial Ecosystems

Until recently,[22] the effects of enhanced solar UV-B radiation on crops and natural terrestrial plant species were generally considered to be negative. Reduction of biomass, reduced yields of crops and disturbed photosynthetic processes have been reported as a result of indoor UV-B studies.

Relatively few studies report on UV-B effects on native plant species in their natural ecosystem. Negative effects of enhanced UV-B radiation simulating realistic scenarios of stratospheric ozone depletion tend to be less than predicted from greenhouse studies. In two reviews,[21,22] direct damaging reactions of plants and crops were stressed. Reduced plant growth under enhanced UV-B was expressed as a reduction of plant height, plant dry weight and leaf area. In contrast with these earlier views, evidence grows that enhanced UV-B does not so much directly affect plant growth and primary production of natural ecosystems,[5,18-20,24] but rather indirectly by UV-B effects on the phenylpropanoid pathway and on plant hormones.

Indirect UV-B Effects on Terrestrial Ecosystems Related to Products of the Phenylpropanoid Pathway, Signal Transduction and Chemical Defence

Solar UV-B radiation is known to stimulate the enzymes PAL and CHS and other branch-point enzymes of the phenylpropanoid pathway.[6,26] PAL catalyzes the transformation of phenylalanine into *trans*-cinnamic acid, which may lead to the formation of complex phenolic compounds such as flavonoids, tannins and lignin.

Induction of UV-B-absorbing *flavonoid* synthesis is primarily considered to function as a UV screen. Flavonoids accumulate particularly in vacuoles of epidermal cells,[25] reducing UV-B levels in the chloroplasts of mesophyll cells. Flavonoids also serve various other functions in plants: signal transduction, flower colour, nectar guides and controlling plant–plant and plant–microorganism interactions (Figure 10). Flavones and flavonols are secreted by legume roots and

Figure 8 (a) UV-B supplementation (middle and right) and filtration systems (left) installed in a dune grassland ecosystem, Heemskerk, The Netherlands (see box for further details)

8 (a) Philips TL 12/40 lamps in wooden frames have been installed over the dune grassland vegetation. Frames are 1.80×2.40 m. There are 5 fluorescent tubes spaced at 40 cm, installed at a height of 1.50 m above the dune soil. Ambient UV-B treatments and enhanced UV-B treatments are adjacent (pairwise) and separated above-ground by a vertical sheet of Mylar foil. Below-ground adjacent treatments are separated by strips of 40 cm deep stainless steel.

In the foreground Open Top Chambers ('white chimneys') are used to study the response of dune grassland species to atmospheric carbon dioxide enrichment. Carbon dioxide enrichment research forms part of the UVECOS project.

The UV-B filtration system consists of a PVC pipeline frame with plastified mash wire carrying 4 overlapping strips of cellulose diacetate foil, transmitting radiation >290 nm (ambient UV-B) or mylar foil excluding radiation with a wavelength >315 nm (UV-B) and transmitting UV-A (below ambient UV-B). The overlap zone of the filter strips is 5 cm. Natural precipitation is continuously measured nearby the UV-filter systems. If necessary the dune vegetation is watered with tap water to compensate for the rainfall intercepted by the filters.

The UV-B filter systems are 130×145 cm wide and only vegetation of the inner 80×80 cm is studied to avoid edge effects, *i.e.* the incidence of unfiltered sunlight at the edge of the plot. Condensation on the lower side of the filters will reduce transmittance of solar radiation in the morning hours, particularly in humid spring and autumn periods.

The filter systems appear to be wind and storm proof. There are also control plots, which are not covered with polyester foil, to assess the impact of rainfall interception by the foil. To assess 'filter effects' on the vegetation, *i.e.* effects due to the presence of the filter above the plants or vegetation, always 120×120 cm control areas or plots should be studied. In the summer months June, July, August we could not measure a significant increase of temperature beneath the filters in the dune grassland. In the autumn (October, November) condensation on the lower side of the filters may reduce transmittance of the filters, particularly in the morning hours.

The outdoor filter systems allow a comparison of plant response to ambient solar UV-B (about $5\,\mathrm{kJ\,m^{-2}\,day^{-1}}$ UV-B_{BE}) and below-ambient solar UV-B radiation (less than $1\,\mathrm{kJ\,m^{-2}\,day^{-1}}$ UV-B_{BE} beneath the plastic filters.

Figure 8 (b) Close up of the UV-B supplementation system with the UV lamps used in the dune grassland ecosystem in Heemskerk (see box for further details)

> **8 (b)** The lamps are enveloped by UV-B-absorbing perspex. Below the lamps is a strip of UV-B-transparent perspex (acrylate), on which either a strip of cellulose acetate foil (enhanced UV-B) is placed or a strip of mylar foil (ambient UV-B). Lamps of both the enhanced UV-B and ambient UV-B treatment burn. In this way, UV-A radiation emitted by the lamps does not differ between the ambient and enhanced UV-B radiation treatment. A more gradual increase of supplemented UV-B radiation can be reached by a stepwise increase of the number of lamps burning in a rack. There is a stepwise increase and decrease of the numbers of lamps functioning, with all lamps switched on around noon. The burning period of the lamps is further varied with the yearly course of global solar radiation, with a maximum burning period around June. The biologically effective UV-B dose in the above-ambient UV-B simulates a 15% stratospheric ozone depletion

regulate gene expression in nodulating N_2 fixing *Rhizobium* bacteria. Also, there is evidence that UV-B-induced flavonoids affect vesicular arbuscular mycorrhiza infection of sugar maple[32] and the dune grassland species *Carex arenaria* and *Calamagrostis epigejos*. In these studies a negative effect of enhanced UV-B radiation on mycorrhizal infection was found.[33]

Polyphenolic tannins affect the palatability and digestibility of plants. The astringency of tannins in plants is repellent to insects, reptiles, birds and higher animals. Tannins also form bonds with peptides and proteins. This bonding prevents plant proteins being attacked by animal digestive enzymes.[34,35] An increased content of tannins relating to enhanced UV-B affects herbivory and

---

[32] J. N. Klironomos and M. F. Allen, *Funct. Ecol.*, 1995, **9**, 923.
[33] J. W. M. van de Staaij, J. Rozema and R. Aerts, in *Stratospheric Ozone Depletion: Effects of Enhanced UV-B on Terrestrial Ecosystems*, ed. J. Rozema, Backhuys, Leiden, 1999, pp. 159–172.
[34] N. D. Paul, M. S. Rasanayagam, S. A. Moody, P. E. Hatcher and P. G. Ayres, *Plant Ecol.*, 1997, **128**, 296.
[35] N. Paul, T. V. Callaghan, S. Moody, D. Gwynn-Jones, U. Johanson and C. Gehrke, in *Stratospheric Ozone Depletion: Effects of Enhanced UV-B on Terrestrial Ecosystems*, ed. J. Rozema, Backhuys, Leiden, 1999, pp. 117–134.

Figure 9 Mini UV-B lamp system installed over *Deschampsia antarctica* at a north faced slope of rocks at Léonie Island (Antarctic peninsula). In the background is the polar ocean with icebergs and glaciers at the horizon (see box for further details)

> **9** The mini UV-B supplementation system was also installed on other sites to expose Antarctic mosses, lichens and terrestrial algae to enhanced UV-B radiation. A battery supplies the 15 Watt fluorescent tube with electricity. The tube and armature are held within a four-pod stainless steel frame. Above and aside the fluorescent tube and armature are covered by UV-B absorbing perspex. Underneath the tubes strips of cellulose acetate or mylar foil are placed on UV and PAR transparent perspex.
>
> Currently these mini UV-B lamp systems are also applied in a Dutch coastal dune area to expose the dune moss *Tortula ruraliformis* to enhanced UV-B.

other plant–animal relationships (Figure 10).

The physical and chemical stability and the resistance to microbial breakdown of complex polyphenolics (such as lignin) in plant litter contributes considerably to the long-term storage of organic carbon in terrestrial ecosystems.[1,2,5,13,35] The polyphenolic compound lignin determines to a large extent the resistance of dead organic plant matter to degradation by microorganisms. Lignin contributes to the persistence of peat and humus in soils. Thereby, complex polyphenolics help to establish large, long-lasting sinks of carbon in terrestrial ecosystems. The established relative homeostasis between oxygen and carbon dioxide in the current atmosphere[1,2] may partly be based on the presence of recalcitrant polyphenolics in terrestrial soil biota.[12] In some recent studies,[14,15,35] enhanced solar UV-B, simulating about 15% stratospheric ozone depletion, resulted in an increased content of tannins and lignin in terrestrial plants.

Enhanced Solar UV-B and Interactions between Terrestrial Higher Plants and Microorganisms

Litter Decomposition and Higher Plant–Vesicular Arbuscular Mycorrhizae Symbiosis. Enhanced solar UV-B radiation may affect litter decomposition in

Figure 10 Diagram summarizing the direct and indirect effects of solar UV-B radiation on ecosystem structure and functions (see box for further details)

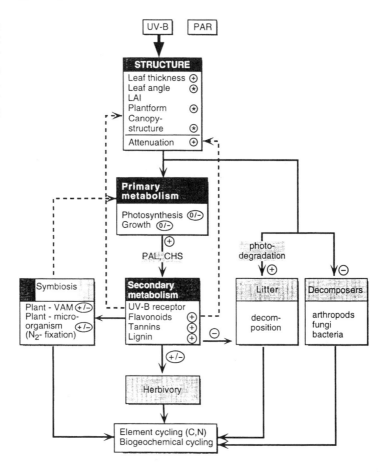

10 Broken lines indicate UV-B-regulated feedback loops, controlling the penetration of solar radiation into the plant canopy and plant tissue. LAI = leaf area index, PAL = phenylalanine ammonia lyase, CHS = chalcone synthase. The blocks with a dark tone represent autotrophic parts of the ecosystem, *i.e.* the mosses, ferns and higher plants, and blocks with a lighter tone indicate heterotrophic parts, such as fungi and microorganisms. +, 0 or − within circles indicate negative, neutral or positive effects, *e.g.* solar UV-B radiation induces activity of the enzymes PAL and CHS: an increased plant content of lignin (positive effect) and this may lead to a decreased rate of litter decomposition (negative effect)

several ways. Changes in plant chemistry may occur through induction by UV-B of PAL and CHS. The altered content of phenolic compounds in leaf litter may indirectly affect litter decomposition. Phenolic compounds reduce the growth of microorganisms, including fungi.

Alternatively, the breakdown of litter may be accelerated by the photochemical action of UV-B. This direct effect of solar UV-B radiation is called photodegradation. In ecosystems with an open canopy and a low leaf area index (LAI), *e.g.* tundra, dry grassland, savanna and deserts, photodegradation may be responsible for rapid decomposition of plant litter since solar UV-B radiation may penetrate into the canopy and litter layer.[13,19] In forests (with a closed canopy and larger LAI) there will be much stronger attenuation of solar UV-B radiation and the photodegradative effect of UV-B on litter decomposition will be small. Experimental and theoretical evidence for the significance of photodegradation of litter in terrestrial ecosystems by solar UV-B has also been provided.[15] Assuming an open canopy (low LAI, erectophilous leaves), solar UV-B may directly irradiate decomposing organisms such as fungi, bacteria and the soil fauna. Elevated solar UV-B may inhibit fungal decomposers and thus reduce the rate of litter decomposition. This has been discussed by Paul et al.[34,35]

In previous studies, growth and photosynthesis of many dune grassland species appeared insensitive to enhanced solar UV-B radiation.[23,36] Dead plant material (litter) originating from *Calamagrostis* plants grown under elevated solar UV-B showed an increased content of lignin and decayed more slowly than plant material grown under ambient UV-B.[13,15] So, indirectly enhanced UV-B may tend to reduce the rate of litter decomposition. Photodegradation by UV-B of dissolved organic matter (DOM) in aquatic ecosystems is well known.[37] Photodegradation of the DOM by UV-B also increases the vertical penetration of solar UV-B into the water column.[38] The UV-B-induced breakdown of DOM, the 'yellow substance', represents increased mineralization and this will affect the primary production of the ecosystem, thus representing a major (positive) feedback within an ecosystem, dependent on the flux of the solar UV-B.

Penetration of part of the incident solar UV-B to the litter suggests that this UV-B radiation may also directly affect decomposer organisms (Figure 10). In a laboratory study, negative effects of UV-B on the decomposer community have been reported by Gehrke et al.[14,16]

It is known that mycorrhizal infection of the dune grassland species *Calamagrostis epigejos* may be high.[39] The degree of vesicular arbuscular mycorrhizae (VAM) infection of roots of dune grassland species, which had been above-ground and exposed to enhanced solar UV-B for six years, appeared to be reduced.[33] Evidence is increasing that, through an altered chemical composition of plants under enhanced UV-B, plant–VAM and plant–herbivore relations are affected.

UV-B and Herbivory: UV-B Affecting the Chemical Defence of Plants. As in terrestrial ecosystems, UV-B may cause shifts in trophic relationships in aquatic ecosystems. It has been demonstrated in mesocosm studies that algae and submerged aquatic plants cultivated under enhanced UV-B, and with an increased content of (as yet unknown) secondary metabolites, are being consumed less easily and less efficiently by zooplankton. Remarkably, a

[36] M. Tosserams, G. W. H. Magendans and J. Rozema, *Plant Ecol.*, 1997, **128**, 284.
[37] G. J. Herndl, A. Brugger, S. Hager, E. Kaiser, I. Overnosterer, B. Reitner and D. Slezak, *Plant Ecol.*, 1997, **128**, 42.
[38] D.-P. Häder, *Plant Ecol.*, 1997, **128**, 4.
[39] J. Rozema, W. Arp, J. van Diggelen, M. van Esbroek, R. Broekman and H. Punte, *Acta Bot. Neerl.*, 1986, **35**, 457.

heterotrophic zooplankton species was unable to synthesize UV-B absorbing carotenoids. Carotenoids are produced by autotrophic algae and the zooplankton obtains UV-B absorbing carotenoids by the consumption of (carotenoid containing) algae.[19] This is an example of how solar UV-B affects relationships (feedbacks) between autotrophic and heterotrophic components of ecosystems.

Evidence is growing that UV-B induced chemical changes in plants reduce herbivory. Larvae of the black root fly (Sciaridae) live in the superficial soil layers in the pots in which the dune plant *Arabidopsis thaliana* is cultivated. The larvae feed on plant roots and green plant parts just above the soil surface. The larvae will not often and not continuously be exposed to the supplemented UV-B radiation in the greenhouse. Under the enhanced UV-B radiation the percentage of *Arabidopsis* plants being grazed was reduced. One explanation is that UV-B radiation has directly (negatively) affected the insect larvae. Alternatively, the increased content of UV-B-absorbing compounds in the leaf tissue of *A. thaliana* indicates that the synthesis of flavonoids and possibly other secondary metabolites (tannins?) has been induced by UV-B radiation. Similarly, in another experiment in a controlled environment, *A. thaliana* plants grown under enhanced UV-B appeared to be less infected by aphids than control plants grown without UV-B. While all stages of the life cycle of these aphids (Aphidoidea) in the greenhouse will be aerial, it is difficult to assess the precise UV-B dose received by the aphids. The reduced infection of *A. thaliana* by aphids under enhanced UV-B is marked. This may be caused by UV-B-induced antifeeding (poly)phenolics. Alternatively, direct negative effects of UV-B on the aphids may be involved as well.

A recent study[40] has indicated that the growth of white clover (*Trifolium repens*) may be reduced to a much greater extent than that of ryegrass (*Lolium perenne*) as a result of increased UV-B radiation. This is regarded as important because of the role of this legume species in the pastoral agriculture and the high UV-B levels in New Zealand. Insects reared on a UV-B-sensitive white clover cultivar exposed to high UV-B exhibited only two-thirds the growth of insects reared on this cultivar without UV-B. No UV-B effect occurred for insects reared on a less UV-B-sensitive cultivar. Changes in performance were due primarily to reduced efficiencies with which food was converted to insect biomass, rather than to reduced food consumption rates. Cyanogenesis in the UV-B-sensitive cultivar may be enhanced by growth under high UV-B radiation. The UV-B-insensitive populations of *Trifolium* have proven in tests to be acyanogenic. This may be part of the explanation for the observed effects on insect growth rates.[40] Cyanogenesis in *Trifolium* relates to cyanogenic glycosides, *e.g.* linamarin, a nitrogen-containing secondary metabolite. The above case confirms that possibly UV-B-induced secondary metabolites may change the relationship between *Trifolium repens* and herbivorous insects.

7 Conclusions and Outlook

There is important physiological and biochemical knowledge of the effects of enhanced solar UV-B radiation on terrestrial plants, mainly based on laboratory

---

[40] B. D. Campbell, R. W. Hofmann and C. L. Hunt, in *Stratospheric Ozone Depletion: Effects of Enhanced UV-B on Terrestrial Ecosystems*, ed. J. Rozema, Backhuys, Leiden, 1999, pp. 227–250.

and climate room studies. Enhanced solar UV-B may be an environmental stress factor, disturbing membrane functioning and causing DNA damage, leading to reduced biomass and yield. In field studies on natural ecosystems, indirect effects of enhanced UV-B radiation on plant secondary metabolites and morphogenetic parameters (canopy architecture, competitive relationships between plant species, plant–animal relationships and litter decomposition) appear to be as important as direct effects of UV-B on plant growth and primary production.

There is little evidence that enhanced solar UV-B will markedly depress plant growth and primary production of terrestrial ecosystems. Yet, ambient and above ambient levels of UV-B represent (sometimes ephemeral) environmental stress to plants. Plant exposure to UV-B generally induces production of UV-B-absorbing secondary metabolites. When plant growth is reduced under (enhanced) UV-B, how does this relate to costs involved in the production of secondary metabolites, in the repair of DNA damage or membrane damage, or in scavenging of radicals? At the moment, we can only say that the metabolic costs to the plant of various UV-B protection mechanisms are not known.

Recent field studies have demonstrated that crop plants (*Pisum sativum* and *Oryza sativa*), which were reported earlier to be sensitive to UV-B in growth chamber studies, were not much affected by elevated UV-B. Also in ecosystems experiments the effects of enhanced UV-B on the growth and primary production are relatively small.[19,20,41] Apparently, under field conditions these plants were able to effectively mitigate any UV-B damage.

UV-B has been shown to induce plant morphogenetic effects occurring both in controlled-environment studies and in field ecosystem studies. These effects may have far-reaching ecological consequences, such as shifts in competitive relationships between plant species, but only one study has thus far demonstrated this and only a few studies are under way to further evaluate these phenomena.

Secondary metabolites in part induced by solar UV-B have played an important role in the evolution of land plants, not only as UV filters but also in many other ecological relationships. Induction of increased secondary metabolites by enhanced solar UV-B has been demonstrated in some studies and this is affecting not only changes in litter decomposition, carbon cycling and herbivory in terrestrial ecosystems, but also symbiotic relationships between higher plants and microorganisms. Therefore, solar UV-B radiation is not just a potential environmental stress for plants, but also affects major aspects of the structure and functioning of terrestrial biota.

Recent assessments[31] demonstrate that there are no signs of stratospheric ozone recovery and an Antarctic stratospheric ozone hole may exist until 2050 and beyond. It is foreseen that ecosystem studies of the effects of enhanced UV-B will reveal changes in trophic relationships within the terrestrial biota.

8 Acknowledgements

Research funding was by the Dutch National Research Programme on Global Air Pollution and Climate Change (project number 850022) and by EU contracts

---

[41] A. M. C. Oudejans, A. Nijssen, J. Huls and J. Rozema, *Plant Ecol.*, 2000, in press.

ENV4-CT96-0208 and ENV4-CT97-0580, which is gratefully acknowledged. We thank Dr. Cees de Vries, Dr. Harm Snater, Dr. Rienk Slings, Mr. Hidde Posthuma and the late Mr. Evert Koet for the permission and support of the UV-B and CO_2 experiments in Heemskerk. The research described on the development and installation of UV mini lamp systems in particular (project number 751.499.06) was financially made possible by the Netherlands AntArctica Programme (NAAP) of The Netherlands Geosciences Foundation of NWO, which is gratefully acknowledged.

Sunlight, Skin Cancer and Ozone Depletion

BRIAN L. DIFFEY

1 Introduction

The sun is responsible for the development and continued existence of life on Earth. We are warmed by the sun's infrared rays and we can see with eyes that respond to the visible part of the sun's spectrum. More importantly, visible light is essential for photosynthesis, the process whereby plants, necessary for our nutrition, derive their energy. Besides serving as the ultimate source of his food and his energy, sunlight also acts on man to alter his chemical composition, control the rate of his maturation and drive his biological rhythms. However, the ultraviolet (UV) component, which comprises approximately 5% of terrestrial solar radiation, is largely responsible for the deleterious effects associated with sun exposure. In particular, the shorter wavelengths of terrestrial UV radiation (UV-B waveband: 290–320 nm) are especially damaging. A major influence on the spectral irradiance of this waveband reaching the Earth's surface is stratospheric ozone.

In 1974, Molina and Rowland—who along with Paul Crutzen were awarded the 1995 Nobel Prize for Chemistry—predicted that man-made chlorine compounds released at ground level would diffuse into the upper atmosphere and destroy the ozone resident there.[1] It was not for another 10 years that scientists from the British Antarctic Survey[2] showed that each year since the mid-1970s there had been an unexpected decrease in the abundance of springtime ozone over the Antarctic—the so-called *ozone hole*. Since then, there has been increasing public concern about just what is going on in the skies above us, fuelled by media speculation of skin cancer epidemics.

In principle, a reduction in ozone levels could result in an increase in the incidence of harmful effects on health that are known to be caused by the UV-B radiation in sunlight. The principal deleterious effects of UV-B radiation are:

[1] M. Molina and F. S. Rowland, *Nature*, 1974, **249**, 810.
[2] J. C. Farman, B. G. Gardiner and J. D. Shanklin, *Nature*, 1985, **315**, 207.

Issues in Environmental Science and Technology No. 14
Causes and Environmental Implications of Increased UV-B Radiation

2 Trends in Atmospheric Ozone and Ambient Ultraviolet Radiation

Significant global scale decreases in total ozone have been occurring since the late 1970s, and the loss of ozone in the northern hemisphere is now proceeding with a rate of loss over mid-latitudes (30–50°N) seen in winter and early spring of about 6% per decade. The loss in summer months, when UV levels are much higher and people are exposed more frequently to the sun, is less at about 3% per decade[3] (see also Chapter 1).

Calculations for the northern hemisphere based on the measured ozone trends for the period 1979–1992 indicate that, all other factors being constant, the terrestrial biologically active ultraviolet radiation (which lies mainly within the UV-B waveband) should have increased by less than 1% per decade at 15°N to about 3% per decade at 30–40°N and 5% per decade at 50–60°N.[3] Predicted changes in ozone column and ambient UV in northern Europe are illustrated in Figure 1.

Paradoxically, these predictions have not generally been borne out by ground-based UV monitoring programmes.[4] Reasons offered to account for this apparent discrepancy include the limited period of most UV monitoring networks, the accuracy of instrument calibration and long-term stability of monitoring equipment, the year-to-year fluctuations in cloud cover,[5] and an increase in ozone and aerosols present in the lower atmosphere due to pollution.[6] Despite the record low in total ozone which occurred in the northern winters of 1992 and 1993,[7] measurements in the Austrian Alps showed no significant increase of cumulative erythemal exposure compared with a reference series of measurements obtained between 1981 and 1988.[8] These issues are dealt with in more detail in the chapter by Webb.

In the southern hemisphere the influence of Antarctic ozone depletion on ambient UV-B in Melbourne (latitude 38°S) has been reported.[9] Continuous monitoring of ambient UV-B showed that the levels recorded in February 1991 were 37% and 27% higher than for the same month in 1990 and 1989, respectively. February 1991 had the lowest ozone values recorded for this period, but also very low cloud cover compared with recent years. These two important factors reinforce each other and illustrate the difficulty of separating the effects of

---

[3] S. Madronich, R. L. McKenzie, L. O. Bjorn and M. M. Caldwell, *J. Photochem Photobiol.*, 1998, **46**, 5.
[4] B. L. Diffey (ed.), *Measurement and Trends of Terrestrial UV-B Radiation in Europe*, Organizzazione Editoriale Medico Farmaceutica, Milan, 1996.
[5] J. E. Frederick and C. Erlick, *Photochem. Photobiol*, 1995, **62**, 476.
[6] C. Bruhl and P. J. Crutzen, *Geophys. Res. Lett.*, 1989, **16**, 703.
[7] R. D. Bojkov, C. S. Zerefos, D. S. Balis, I. C. Ziomas and A. F. Bais, *Geophys. Res. Lett.*, 1993, **13**, 1351.
[8] M. Blumthaler and W. Ambach, *Lancet*, 1994, **343**, 303.
[9] C. R. Roy and H. P. Gies, in *Proceedings of the 8th International Congress of the International Radiation Protection Association*, 1992, vol. 1, pp. 759–762.

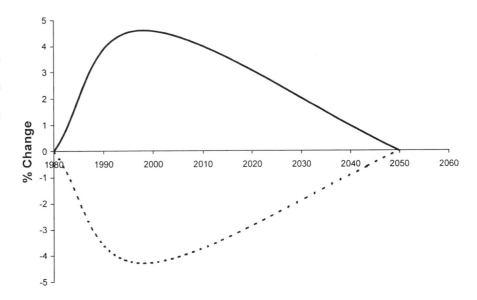

Figure 1 Predicted changes in ozone column (broken curve) and ambient UV (solid curve) in northern Europe over the period 1980–2050 (Adapted from Madronich et al.[3])

ozone depletion from climate on ambient UV-B. A further example of this confounding is that ground-based measurements showed that summertime sunburning UV radiation in the southern hemisphere exceeded those at comparable latitudes of the northern hemisphere by up to 40%,[10] whereas corresponding satellite-derived estimates yield only 10–15% differences.[4] Atmospheric pollution may be one factor in this discrepancy.

So, whilst there is unequivocal evidence concerning stratospheric ozone depletion, we cannot be sure, as yet, whether this depletion is accompanied by increases in terrestrial UV radiation. This does not mean that no systematic trend exists, simply that the 95% confidence interval on estimated trends are likely to encompass zero. Measurement data from the UK National Radiological Protection Board[11] confirm this conclusion (Figure 2). See also the chapter by Webb.

3 Human Exposure to Solar Ultraviolet Radiation

The solar ultraviolet radiation to which an individual is exposed depends upon:[12]

- Ambient solar ultraviolet radiation
- The fraction of ambient exposure received on different anatomical sites
- Behaviour and time spent outdoors

The UV dose absorbed by the skin is further modified by the use of photoprotective agents such as hats, clothing and sunscreens.

Maximum daily ambient ultraviolet levels under clear summer skies are about

---

[10] G. Seckmeyer, B. Mayer, G. Bernhard, R. L. McKenzie, P. V. Johnston, M. Kotkamp, C. R. Booth, T. Lucas and T. Mestechikina, *Geophys. Res. Lett.*, 1995, **22**, 1889.
[11] C. M. H. Driscoll, personal communication, 1999.
[12] B. L. Diffey, in *Photodermatology*, ed. J. L. M. Hawk, Arnold, London, 1999, pp. 5–24.

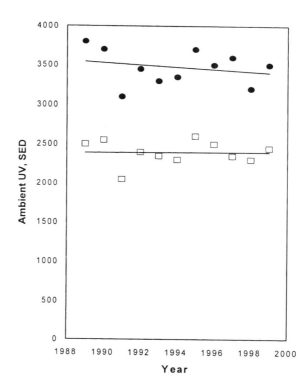

Figure 2 Ambient erythemal UV measured at Chilton, UK (latitude 51.5°N) (Courtesy of Dr C. M. H. Driscoll, NRPB)

70 SED in the tropics, 60 SED at mid-latitudes approximating to those of southern Europe and 45 SED for UK latitudes.[13] The SED (standard erythema dose) is a measure of erythemal UV radiation;[14] it requires an exposure of about 1.5 SED to produce just perceptible reddening of skin (erythema) in the unacclimatized skin of sun-sensitive individuals who burn easily and never tan (skin type I),[15] about 2 SED in subjects who burn easily but tan minimally (skin type II) and 3 SED in subjects who will burn but tan readily (skin type III).[16] In the British population, about 11, 30 and 31% of people are of skin types I, II and III, respectively.[17]

Estimates of personal exposure are normally obtained by direct measurement

[13] C. R. Roy, H. P. Gies and S. Toomey, *Cancer Forum*, 1996, **20**, 173.
[14] CIE Standard, CIE S 007/E-1998, Commission Internationale de l'Éclairage, Vienna, 1998.
[15] J. Lock-Andersen, H. C. Wulf and N. N. Mortensen, in *Landmarks in Photobiology*, eds. H. Hönigsmann, R. M. Knobler, F. Trautinger and G. Jori, Organizzazione Editoriale Medico Farmaceutica, Milan, 1998, pp. 315–317.
[16] M. A. Weinstock, *J. Clin. Epidemiol.*, 1992, **45**, 547.
[17] HEA, *Sunscreens and the Consumer*, Health Education Authority, London, 1996.

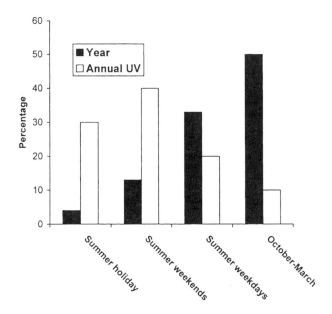

Figure 3 How British adults get their sun exposure. A 2-week summer holiday (4% of the year) contributes 30% of the annual dose of an adult indoor worker, and so on

using UV-sensitive film badges.[18] The results obtained from a number of studies in northern Europe[19-25] indicate that indoor workers receive an annual exposure of around 200 SED, mainly from weekend and vacational exposure, and principally to the hands, forearms and face. This value is approximately 5% of the total ambient available. Children have a greater opportunity for outdoor exposure and receive an annual dose in the UK of around 300 SED.

Outdoor workers at the same latitudes receive about 2–3 times these exposure doses,[20,24] whilst film badge studies[26] on three groups of outdoor workers on the Sunshine Coast in Queensland (27°S) suggest that annual exposures would be considerably higher—certainly in excess of 1000 SED per year.

For indoor workers the annual exposure associated with occupation (travelling to and from work, going outside at lunchtime) is about 40 SED, about 100 SED is contributed by weekend exposure and the remaining 60 SED from vacational exposure (Figure 3). In the case of children, 'occupational' exposure (playtime and lunchtime exposure) may be about 60 SED, recreational about 180 SED (because

[18] B. L. Diffey, in *Radiation Measurement in Photobiology*, ed. B. L. Diffey, Academic Press, London, 1989, pp. 135–139.
[19] A. V. J. Challoner, D. Corless, A. Davis, G. H. W. Deane, B. L. Diffey, S. P. Gupta, and I. A. Magnus, *Clin. Exp. Dermatol.*, 1976, **1**, 175.
[20] O. Larkö and B. L. Diffey, *Clin. Exp. Dermatol.*, 1983, **8**, 279.
[21] J. F. Leach, V. E. McLeod, A. R. Pingstone, A. Davis and G. H. W. Deane, *Clin. Exp. Dermatol.*, 1978, **3**, 77.
[22] A. A. Schothorst, H. Slaper, R. Schouten and D. Suurmond, *Photodermatology*, 1985, **2**, 213.
[23] H. Slaper, *PhD thesis*, University of Utrecht, The Netherlands, 1987.
[24] A. R. Webb, *PhD thesis*, University of Nottingham, UK, 1985.
[25] P. Knuschke and J. Barth, *J. Photochem. Photobiol.*, 1996, **36**, 77.
[26] H. P. Gies, C. R. Roy, S. Toomey, R. MacLennan and M. Watson, *Photochem. Photobiol.*, 1995, **62**, 1015.

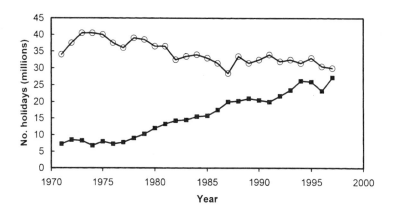

Figure 4 Domestic (open circles) and overseas (closed squares) holidays taken by British residents (data from the Office for National Statistics)

children are at school for only about 190 days per year) and vacation with parents giving about 60 SED. It must be stressed, however, that there will be large variations in the annual exposure doses received by individuals within a given population group, depending upon propensity for outdoor activities.[27]

Behaviour can be equally, or more, important than ambient UV on an individual's sun exposure.[28] The daily ambient erythemal UV in the UK shows a clear sky summer to winter ratio of about 40:1, with day-to-day variations as a result of cloud cover. However, population UV exposure will be subject to even greater variation due to differences in individual behaviour. This is illustrated in the results of a recent study[29] which reported the daily summer sun exposure received by children in England (50–55°N) and Queensland, Australia (21–27°S). Whilst the median daily personal exposure in Queensland was twice that received in England, there was a wide overlap between the two distributions. On any one day the daily solar UV exposure of 17% of English children exceeded the Queensland median, and the exposure of 26% of Queensland children was less than the English median.

One important factor that will increase population dose has probably been the growth of overseas holidays (Figure 4). In recent years the most rapid increases in foreign holiday travel have been to long-haul destinations at low-latitude destinations where UV levels are typically high. For example, holiday visits to the USA (where Florida is the most popular destination) increased 15-fold in the 20 years up to 1997. It seems likely therefore that changing patterns of holidaymaking continue to be an important factor tending to increase the overall UV doses received by the UK population and any associated health risks. Coupled with this is the growth in outdoor leisure activities that has continued in the 1990s, with consequential increases in sunlight exposure.[30]

Future social trends remain uncertain but a continuation of such increases would add to the UV exposure experienced by the population and the risks

[27] B. L. Diffey, C. J. Gibson, R. Haylock and A. F. McKinlay, *Br. J. Dermatol.*, 1996, **134**, 1030.
[28] B. L. Diffey, *Eur. J. Dermatol.*, 1996, **6**, 221.
[29] B. L. Diffey and H. P. Gies, *Lancet*, 1998, **351**, 1101.
[30] Office of National Statistics, *Travel Trends: a Report on the 1997 International Passenger Survey*, HMSO, London, 1998.

4 Effects of Ultraviolet Radiation on Skin

The normal responses of skin to UV radiation can be classed under two headings: acute effects and chronic effects. An acute effect is one of rapid onset and generally of short duration, as opposed to a chronic effect which is often of gradual onset and long duration.[31]

Sunburn

Erythema, or redness of the skin due to dilatation of superficial dermal blood vessels, is one of the commonest and most obvious effects of ultraviolet exposure ('sunburn'). Following exposure to solar UV radiation, there is usually a latent period of 2–4 h before erythema develops. Erythema reaches maximum intensity between 8 and 24 h after exposure, but may take several days to resolve completely. If a high enough exposure has occurred, the skin will also become painful and oedematous, and blistering may result.

Tanning

Another consequence of exposure to solar UV (which at present still seems to be socially desirable) is the delayed pigmentation of the skin known as tanning, or melanin pigmentation. Melanin pigmentation of skin is of two types: (i) constitutive—the colour of the skin seen in different races and determined by genetic factors only; and (ii) facultative—the reversible increase in tanning in response to sun exposure.

Hyperplasia

In addition to tanning, the skin is capable of another perhaps even more important adaptive response which limits damage from further ultraviolet exposure: epidermal thickening or hyperplasia. This begins to occur around 72 h after exposure, is a result of an increased rate of cell division in the lower epidermis, and eventually results in thickening of both epidermis and stratum corneum (the outermost layer of the epidermis) which persists for several weeks. This adaptive process, unlike tanning, does not depend on a genetic predisposition and is the major factor which protects those who tan poorly in sunlight.

Production of Vitamin D

The only well-established beneficial effect of solar UV on the skin is the production of vitamin D_3. The skin absorbs UV-B radiation in sunlight to convert sterol precursors in the skin, such as 7-dehydrocholesterol, to vitamin

---

[31] B. L. Diffey, in *Environmental Dermatology*, ed. O. Y. Oumeish, Elsevier, New York, 1998, pp. 83–89.

D_3. Vitamin D_3 is further transformed by the liver and kidneys to biologically active metabolites such as 25-hydroxyvitamin D; these metabolites then act on the intestinal mucosa to facilitate calcium absorption, and on bone to facilitate calcium exchange.

Skin Cancer

The three common forms of skin cancer, listed in order of seriousness, are basal cell carcinoma (BCC), squamous cell carcinoma (SCC) and malignant melanoma (MM). Around 90% of skin cancer cases are of the non-melanoma variety (BCC and SCC) with BCCs being approximately four times as common as SCCs. Exposure to ultraviolet radiation (UVR) is considered to be a major etiological factor for all three forms of cancer.[32] For basal cell carcinoma and malignant melanoma, neither the wavelengths involved nor the exposure pattern that results in risk have been established with certainty, whereas for squamous cell carcinoma, both UV-B and UVA are implicated and the major risk factors seem to be cumulative lifetime exposure to UVR and a poor tanning response.

Squamous Cell Cancer. The evidence that exposure to sunlight, even without ozone depletion, is the predominant cause of squamous cell cancer in man is very convincing. These cancers occur almost exclusively on sun-exposed skin such as the face, neck and arms, and the incidence is clearly correlated with geographical latitude, being higher in the more sunny areas of the world.[33] Recent epidemiological studies suggest that sun exposure in the 10 years prior to diagnosis may be important in accounting for individual risk of SCC.[34]

Basal Cell Cancer. The relationship between basal cell carcinoma and sunlight is less compelling, but the evidence is sufficiently strong to consider it also to be a consequence of exposure to sunlight. Whilst SCC is strongly related to cumulative lifetime exposure to sunlight, this relationship is not so convincing for BCC,[35,36] and it may be that sun exposure in childhood and adolescence may be critical periods for establishing adult risk for BCC.[35]

Malignant Melanoma. During the past 40 years or so there has been an increase in the incidence of malignant melanoma in white populations in several countries. There exists an inverse relationship between latitude and melanoma incidence and this has been taken as evidence for a possible role of sunlight as a cause of malignant melanoma. However, this pattern is not always consistent. In Europe, for example, the incidence and the mortality rates in Scandinavia are considerably higher than those in Mediterranean countries. This inconsistency may reflect

---

[32] B. K. Armstrong and A. Kricker, *Dermatoepidemiology*, 1995, **13**, 583.
[33] A. Kricker, B. K. Armstrong and D. R. English, *Cancer Causes & Control*, 1994, **5**, 367.
[34] R. P. Gallagher, G. B. Hill, C. D. Bajdik, S. Fincham, A. J. Coldman, D. I. McLean and W. J. Threlfall, *Arch. Dermatol.*, 1995, **131**, 164.
[35] R. P. Gallagher, G. B. Hill, C. D. Bajdik, S. Fincham, A. J. Coldman, D. I. McLean and W. J. Threlfall, *Arch. Dermatol.*, 1995, **131**, 157.
[36] A. Kricker, B. K. Armstrong, D. R. English and P. J. Heenan, *Int. J. Cancer*, 1995, **60**, 489.

ethnic differences in constitutional factors and customs. Also, the unexpectedly low incidence in outdoor workers, the sex and age distribution, and the anatomical distribution have pointed to a more complex association.[37]

There is now growing evidence that intermittent sun exposure—mainly from recreational activities—rather than cumulative or chronic exposure associated with occupation is associated with increased risk of developing malignant melanoma. Several studies have established a history of sunburn as an important risk factor for melanoma development, although in these studies a potential for recall bias exists. Migration studies have led to the suggestion that sun exposure in childhood is a particularly critical period in terms of melanoma risk.

5 Risk Analysis of Human Skin Cancer

Estimates of the risk of inducing skin cancer from exposure to ultraviolet radiation demands knowledge of dose–response relationships and the relative effectiveness of different wavelengths (known as an *action spectrum*) in sunlight in causing skin cancer. Data on dose–response relationships and action spectra are available to some extent to allow quantitative estimates of the effects of ozone depletion on non-melanoma skin cancer (NMSC) incidence. These data remain unknown for malignant melanoma and so it is unwise to make predictions about the consequences of ozone depletion on this disease. It cannot be assumed, however, that because knowledge does not allow risk estimates to be made for malignant melanoma, ozone depletion will have no effect on the incidence of this disease.

Dose–Response Relationships

Application of multivariate analysis to population-based epidemiology of NMSC has shown that, for a group of subjects with a given genetic susceptibility, age and environmental ultraviolet exposure are the two most important factors in determining the relative risk. Other epidemiological studies have confirmed these findings, and this has led to a simple power law relationship which expresses the risk in terms of these factors:

$$\text{Risk} \approx (\text{annual UV dose})^\beta \, (\text{age})^\alpha$$

The symbols α and β are numerical constants associated with the specific type of NMSC and are normally derived from surveys of skin cancer incidence and ultraviolet climatology. For BCC, exemplary values of α and β are 3.2 and 1.7, respectively;[38] for SCC the corresponding values are 5.1 and 2.3.[38]

This equation is applicable to situations where the annual exposure received by an individual remains unaltered throughout life. In most instances, changes in lifestyle with age mean that the annual UV exposure does not remain constant and so more complex mathematical models of NMSC risk have been developed,[38] where the risk of NMSC at age T is given as:

---

[37] B. K. Armstrong and A. Kricker, *Cancer Surveys*, 1994, **19**, 219.
[38] National Radiological Protection Board, *Health Effects from Ultraviolet Radiation*, Report of an Advisory Group on Non-ionising Radiation, Documents of the NRPB, 1995, vol. 6, no. 2, p. 155.

$$\text{Risk} \approx (\text{cumulative UV dose at age } T)^{\beta-1} \sum_{t=0}^{t=T} (\text{annual dose at age } [T-t])t^{\alpha-\beta}$$

Action Spectrum for Photocarcinogenesis

Clearly an action spectrum for skin cancer can only be obtained from animal experiments. The most extensive investigations to date are those from groups at Utrecht and Philadelphia. These workers exposed a total of about 1100 white hairless mice to 14 different broad-band ultraviolet sources and by a mathematical optimization process derived an action spectrum referred to as the Skin Cancer Utrecht–Philadelphia (SCUP) action spectrum.[39] The SCUP action spectrum is that for skin tumour induction in hairless mice, a species which has a thinner epidermis than humans.

By taking into account differences in the optics of human epidermis and hairless albino mouse epidermis, the experimentally determined action spectrum for tumour induction in mouse skin can be modified to arrive at a postulated action spectrum for human skin cancer.[40] The resulting action spectrum resembles the action spectrum for erythema[41] (Figure 5), suggesting that this action spectrum may be used as a surrogate for human skin cancer.

Risk Calculations

A 1% decrease in ozone will lead to a 1.2–1.4% increase in carcinogenic-effective UV radiation.[42] For every 1% increase in carcinogenic-effective UV-B radiation there will be an approximate 2.3% increase of SCC incidence, with a corresponding figure of approximately 1.7% for BCC (the parameter β referred to above). Combining these we arrive at overall amplification factors, which can be summarized as:

$$1\% \downarrow \text{ in } O_3 \rightarrow 1.4 \times 2.3 = 3.2\% \uparrow \text{SCC}$$
$$\rightarrow 1.4 \times 1.7 = 2.4\% \uparrow \text{BCC}$$

This simple approach can be used to estimate the consequences of a possible, but unlikely, sustained 5% stratospheric ozone depletion in the UK. Figures for skin cancer incidence in the UK are based on 1989 data and are obtained from the Office of Population Census and Surveys for England and Wales, Information and Statistics Division of the NHS Directorate of Information Services for Scotland, and the Northern Ireland Cancer Registry. These sources yield a combined figure of just over 40 000 cases of skin cancer annually in the UK. Of these, about 30 000 are BCC, about 6000 are SCC and the remaining 4000 are

---

[39] F. R. de Gruijl, H. J. C. M. Sterenborg, P. D. Forbes, R. E. Davies, C. Cole, G. Kelfkens, H. van Weelden, H. Slaper and J. C. van der Leun, *Cancer Res.*, 1993, **53**, 53.

[40] F. R. de Gruijl and J. C. van der Leun, *Health Phys.*, 1994, **67**, 319.

[41] A. F. McKinlay, B. L. Diffey and W. F. Passchier, in *Human Exposure to Ultraviolet Radiation: Risks and Regulations*, eds. W. F. Passchier and B. F. M. Bosnjakovic, Elsevier, Amsterdam, 1987, pp. 83–87.

[42] Health Council of the Netherlands, publication no. 1994/05E, Health Council of the Netherlands, The Hague, 1994.

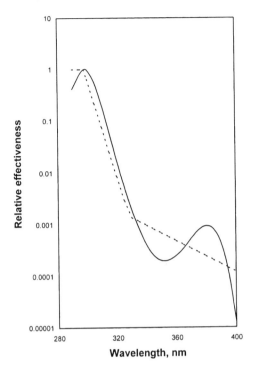

Figure 5 Action spectra for erythema[41] (dashed line) and non-melanoma skin cancer[40] (solid line) in human skin

malignant melanoma. However, because of under-recording of non-melanoma skin cancers (BCC and SCC) in the UK and the rising incidence, the number of these cancers occurring each year is probably closer to double these figures. So, for a sustained 5% ozone depletion, we might expect an increase of 12% and 16% in the incidence of BCC and SCC, respectively, giving an additional 7200 cases of BCC and 2000 cases of SCC each year in the UK.

Whilst this approach may be applicable to future generations who may live under a permanently depleted ozone layer, it is not appropriate to British people alive today. By combining ultraviolet climatological data for the UK with models of human behaviour in order to estimate annual exposure throughout life, and then employing the mathematical models described above, it is possible to estimate the lifetime risk of NMSC. Perturbation of climatological UV-B expected as a consequence of ozone depletion presently occurring is also incorporated to examine the effects of environmental change on skin cancer incidence.[43]

For British adults alive today, ozone depletion continuing indefinitely at current rates is predicted to result in a relatively small (<5%) additional lifetime risk of non-melanoma skin cancer, assuming no changes in climate, time spent outdoors, behaviour or clothing habits.[38] The lifetime risk incurred by today's children, however, is predicted to be 11–16% greater than expected in the absence of ozone depletion. However, if the production and use of substances which deplete ozone are reduced as expected under the provisions of the Montreal

---

[43] B. L. Diffey, *Phys. Med. Biol.*, 1992, **37**, 2267.

Protocol, the increased lifetime risk of skin cancer is likely to be less than these estimates, typically between 4 and 10%.

Recent estimates[44,45] suggest that the increased risk of skin cancer due to ozone depletion would not have been adequately controlled by implementation of the Montreal Protocol alone, but can be achieved through implementation of its later amendments (Copenhagen 1992 and Montreal 1997). These estimates indicate that under the Montreal Amendments, incidences of skin cancer (all types) will peak around the mid-part of the century at additional incidences of about 7 per 100 000. For the UK population of approximately 60 million, this would imply 4200 additional cases of skin cancer per year. Thereafter the increase in disease rates attributable to ozone depletion is expected to return almost to zero by the end of the century; as skin cancer typically results from several decades of UV exposure, the response of the disease follows later than changes in exposure.

The quantitative risk estimates for skin cancer are only valid if all other factors which determine risk, notably human behaviour, remain unchanged. Public health campaigns aimed at encouraging people to reduce their sun exposure by sun avoidance and the use of photoprotective measures, such as sunscreens, clothing and shade, may achieve a reduction in average population UV-B exposures, and presumably skin cancer rates, which could more than offset the adverse effects of ozone depletion.[46] Some preventive measures, however, may not be as effective as commonly believed. The role of sunscreens in the prevention of skin cancer is controversial, with some studies suggesting that sunscreen use is associated with increased risk of malignant melanoma.[47] The reasons for this observation are not known. One of several possibilities is that sunscreen use encourages people to stay in the sun longer, coupled with fact that inadequate amounts of sunscreen are normally applied and/or areas of the body are missed.[48]

Furthermore, the calculated risks imply full compliance with restrictions on the production and consumption of ozone-depleting chemicals throughout the world. If, in the future, compliance does not continue, damage to the ozone layer could be greater than hitherto expected and biological impacts could be more severe.[45]

Another area of future uncertainty relates to the possible effects of greenhouse gas-related climate change. Most attention has been given to the effects of changes in terrestrial UV radiation as a result of stratospheric ozone depletion. What has frequently been overlooked is that greenhouse-gas induced climate change may also independently influence population exposure to sunlight. One reason is that clouds have a large effect on the amount of UV-B reaching the surface and therefore changes in cloud cover would affect exposure.[49] Another factor is that a change to warmer conditions could encourage behaviour

[44] H. Slaper, G. J. M. Velders, J. S. Daniel, F. de Gruijl and J. C. van der Leun, *Nature*, 1996, **384**, 256.
[45] United Nations Environment Programme, *Environmental Effects of Ozone Depletion: 1998 Update*, UNEP, Nairobi, 1998.
[46] D. Hill and J. Boulter, *Cancer Forum*, 1996, **20**, 204.
[47] P. Autier, J.-F. Dor, M. S. Cattaruzza, F. Renard, H. Luther, F. Gentiloni-Silverj, E. Zantedeschi Mezzetti, I. Monjaud, M. Andry, J. F. Osborn and A. R. Grivegne, *J. Natl. Cancer Inst.*, 1998, **90**, 1873.
[48] B. L. Diffey, *Br. Med. J.*, 2000, **320**, 176.
[49] S. Thiel, K. Steiner and H. K. Seidlitz, *Photochem. Photobiol.*, 1997, **65**, 969.

Sunlight, Skin Cancer and Ozone Depletion

associated with greater exposure to the sun and a higher risk of sunburn,[46] and presumably skin cancer. The result might be a marked change in patterns of behaviour that would increase population exposure to sunlight and the health risks associated with it.

Subject Index

Absorbance coefficient, 64–67
Absorption, 21
Absorptivity, 41
Actinometers, 46
Action spectra, 44, 115
   spectrum for hydrogen peroxide, 77
Aerosol, 23, 24
Air quality, 25
Albedo, 21
Ammonia lyase, 87
Ammonium ion, 54
Anchovy, 76
Antarctic region, 8, 9, 15, 20, 23, 100
   ozone hole, 4, 72
Anti-correlation with ozone, 32
Arctic region, 1, 8, 9, 15
Atmospheric model, 13
   chemistry/climate models, 14
Autotrophic organisms, 85

Bacteria, 56
Basal cell cancer, 114
Behaviour, 109
Biogeochemical cycling, 57
Biological weighting function, 69
Biosphere, 18
Black sea model, 47
Broadband radiometers, 29

Calamagrostis epigejos, 99
Calibration, 30
Capillary waveguide, 66
Carbon dioxide, 24, 85, 90
Carbon monoxide, 58, 64

Carbonyl groups, 53
Carbonyl sulfide, 64
Carex arenaria, 99
Catalytic chlorine cycles, 6
   cycles, 4, 23
   reactions, 3
CDOM, *see* Chromophoric dissolved
   organic material
Cellulose diacetate foil, 98
CFCs, *see* Chlorofluorocarbons
Chalcone synthase, 87
Chapman reactions, 3
Chemical actinometry, 66
   defence, 102
Chemistry/climate change, 14
   climate feedback, 13
   climate models, 35
Children, 111
Chlorine monoxide (ClO), 5, 6, 9, 15
Chlorofluorocarbons, 4, 7, 23, 39, 85
Chromophores, 67
Chromophoric dissolved organic
   material (CDOM), 63, 64, 65, 67, 81, 82
   absorption spectra, 41, 48
Cloud, 22
Cod, 76
Commission Internationale
   de'Eclairage, 17
Controlled environment, 95
Copepod, 76
Copper, 55
Cryptophytes, 74

Subject Index

Cuticular waxes, 93
Cyclobutane dimer, 93, 94

Damage and repair processes, 63, 68
Database, 31
Decomposition, 101
Deschampsia antarctica, 95
Diatoms, 72, 74
Diffuse attenuation coefficient, 47
Dimethyl sulfide, 53, 65
Dimethyl sulfoxide, 53
Dinoflagellates, 72, 74
Dissolved organic material (DOM), 40, 62, 63, 102
Dissolved inorganic carbon, 65
Dissolved oxygen, 49
DNA, 67, 72
   damage, 104
   photolyase, 96
   repair, 93, 99
Dobson spectrophotometer, 19
Dose–response relationships, 115
Dosimeter, 19

Einstein unit, 39
Electron transport, 72
Energies of bonds, 39
Energy (E) of a photon, 38
Epidemiological studies, 20
Erythema, 9, 113
Erythemally effective radiation, 20
Evolution, 91
Exhaust emissions, 24
Excision repair, 94
Exposure response curve (ERC), 68
Extraterrestrial solar radiation, 17

Filters, 96
Fish and zooplankton, 71
Flavonoids, 87
Flavonol, 89
Formaldehyde, 58

Global warming, 24
Greenhouse effect, 85
Greenhouse gases, 1, 25, 118

Ground-based observations, 27

Halogens, 24
Herbivory, 103
Holidays, exposure to UV radiation during, 112
Human health, 25, 109–119
Hydrocarbons, 26
Hydrogen peroxide, 50, 64, 67, 71, 77, 80, 81
Hydroxyl radical, 51, 77
Hyperplasia, 113

Indoor workers, 111
Industrialization, 35
Infrared radiation, 1, 3
Inhibition of phytoplankton photosynthesis, 70
Input parameters, 34
Instrument intercomparisons, 30
International Ozone Trends Panel, 6
Iron, 55
Isoflavonoids, 89

Lambert–Beer law, 40
Leisure activities, 112
Lignin, 87

Malignant melanoma, 114
Manganese, 55
Marine bacteria, 75
Melanin pigmentation, 113
Methyl bromide, 7
Middle latitudes, 10
Mie scattering, 21
McMurdo Station, 72
Model atmosphere, 34
Montreal Protocol, 7, 11, 12, 15, 118
Mt Pinatubo, 10, 12, 13
Multi-channel instruments, 29
Mylar foil, 98

Networks, for ozone monitoring, 29
Nitrate, 54, 55
Nitrogen oxides, 24, 26
Non-melanoma skin cancer, 115

Subject Index

Open top chambers, 98
Outdoor workers, 111
Oxygen, 49, 67, 77, 78
Ozone, 17
  absorption coefficient, 18
  depletion, 1–16, 17–36, 90, 108
  hole, 20
  layer, evolution of, 85
  recovery, 35

Palmer Station, 72
PAR, 93
PAR/UV-B ratios, 95
Pathlength, of radiation, 21
Phenylalanine, 87, 97
Phenylpropanoid pathway, 87
Phosphate, 54
Photobleaching, 55, 64, 66–67
Photochemical smog, 26
Photodegradation, 101
Photo-induced polymerisation, 56
Photoinhibitron, 70
Photolysis, 22
Photomorphogenic responses, 95
Photons, 38
Photobleaching, 64, 66, 67
Photorepair, 94
Photosensitised reactions, 42, 45
Photosynthesis, 78, 68, 86
Phytoplankton photosynthesis, 71, 72
Polar stratospheric clouds (PSCs), 6, 9, 10, 15, 23
Polychromatic exposures, 69
Polyphenolics, 87
Pre-industrial times, 34
Proteins, 67

Quality assurance, 31
Quality control, 31
Quanta, 38
Quantum yield, 37, 43, 57, 64–67
Quasi-biennial oscillation, 22

Radiation, 3
Radiative transfer models, 27
Radicals, 50
Rayleigh scattering, 21

Reactive oxygen, 77
Reciprocity, 63, 64
Rhizobium bacteria, 99
Rhode River, 73
Ricketts, 19
Risk analysis, 115

Satellite instruments, 27, 34
Scattering of radiation, 21
Sea–air exchange of chemical species, 54
Seasonal cycles, 33
Secondary metabolites, 104
SED (Standard Erythema Dose), 110
Singlet excited species, 42
  oxygen, 26, 49
Skin cancer, 19, 108
Solar simulator, 70
Solar tracking, 96
Solar zenith angle (SZA), 21
Soot, 24
Spectral radiant flux, 66
Spectral weighting functions, 62
Spectroradiometers, 27
Sporopollenin, 90
Squamous cell cancer, 114
Standards of spectral irradiance, 30
Stratospheric ozone, 2, 78, 85, 107
Sulfur gases, 24
Sunburn, 35, 108
Sunscreens, 118
Sunspot cycle, 22
Superoxide anion, 50, 66
Superoxide dismutase (SOD), 94
Supersonic aircraft, 19
Surface microlayer, 54
Survival curve, 68

Tanning, 113
Tannins, 87
Terrestrial biota, 86, 95
Total Ozone Mapping Spectrometer (TOMS), 7, 58
Trace metals, 55
Trends in UV-B levels, 32, 108
Triplet excited species, 42

Subject Index

Tropospheric ozone, 1, 25

Uncertainties, in UV data, 30
UV forecasting, 35
UV index, 35
UV lamps, 70
UV monitoring, 108
UV screens, 97
UV supplementation, 100

Vertical mixing, 48, 81

Vesicular arbuscular mycorrhiza, 99, 102
Vitamin D_3, 113

Weather systems, 23
Weddell–Scotia Confluence, 72

Xenon arc, 70

Zooplankton, 76, 78